# Ecological Studies

Analysis and Synthesis

Edited by

J. Jacobs, München · O. L. Lange, Würzburg

J. S. Olson, Oak Ridge · W. Wieser, Innsbruck

Volume 2

# Integrated Experimental Ecology

Methods and Results of
Ecosystem Research in the German Solling Project

Edited by

## H. Ellenberg

With 53 Figures

Springer-Verlag New York · Heidelberg · Berlin 1971

ISBN 0-387-05074-4 Springer-Verlag New York Heidelberg Berlin
ISBN 3-540-05074-4 Springer-Verlag Berlin Heidelberg New York

# Foreword

In the first volume of this series (edited by D. E. REICHLE) an international panel of experts treated various aspects of ecosystems research, using deciduous forests of the temperate zone as an example. That collection of articles gave a general view of the extent of our present knowledge and established a conceptual framework for the analysis of an ecosystem. The present volume carries on from there, although it serves a different purpose. It is a report on their methods and their experiences by the members of a research group working as a team.

Under the direction of Prof. ELLENBERG, one of the pilot projects of the International Biological Program has been begun in the Solling, a forest and grassland area near Göttingen, West Germany. Here scientists representing a variety of disciplines — meteorology, soil science, hydrology, botany, zoology, microbiology, agriculture and forestry — got together to ascertain the practical possibilities of the analysis of ecosystems. The research, which began in 1966, is still going on. Nevertheless, there are good reasons for making a report on experience so far.

People are becoming increasingly aware that our very existence depends upon an exact study of the environment. All over the world research groups are being formed to gather quantitative data about the functioning of ecosystems. The main goal is to understand the biological, climatic and edaphic bases of biological productivity and thus to learn to evaluate the capacity of the human environment and its ability to resist insult from an economic as well from a cultural standpoint. This second volume of *Ecological Studies*, where the accent is very much on methods, should serve as a practical guide for any such future studies. In addition, it should give all readers some insight into the possibilities of team work on the grand scale in experimental environmental research.

The editors feel that the particular value of this volume is that it is concerned with cooperative research actually going on now. This means that some of the results will be subject to further integration in the future. Not until these studies in the Solling are completed will it be possible to include in this series a final synthesis of the results obtained.

The Series Editors

# Contents

*Part 1: Primary Production*

## A. Measurement of $CO_2$ Gas-Exchange and Transpiration in the Beech (Fagus silvatica L.). By O. L. Lange and E.-D. Schulze

## O. Food and Energy Turnover of Phytophagous and Predatory Arthropods.
— Methods Used to Study Energy Flow. By W. FUNKE and G. WEIDE-MANN

*Part 3: Environmental Conditions*

**S. The Measurement of Climatic Elements which Determine Production in Various Plant Stands.** — Methods and Preliminary Results. By O. KIESE

**T. The Characterization of the Woodland Light Climate.** By W. EBER

*Part 4: Range of Validity of the Results*

**Y. Phenological Comparisons of the Forest Research Area in the Solling with Similar Forests in Other Mountain Ranges.** By F. K. HARTMANN

**Z. Results of a Grassland Mapping in the High Solling.** By B. SPEIDEL

*Editorial Note:*

Several Research Projects are not yet available for publication, mainly:

*In part 1:*

Inventory of Plant and Animal Species of all Research Plots. By many collaborators
Maximum Plant Production on Agricultural Crop Fields. By K. BAEUMER

# Contributors

AHRENS, E., Dr., Institut für Bodenkunde und Waldernährung der Universität, Göttingen

BABEL, U., Dr., Institut für Bodenkunde der Universität, Stuttgart

BENECKE, P., Dr., Institut für Bodenkunde und Waldernährung der Universität, Göttingen

BENNERT, W., Institut für Angewandte Botanik der Technischen Universität, Berlin

BORNKAMM, R., Prof. Dr., Institut für Angewandte Botanik der Technischen Universität, Berlin

BOMMER, D., Prof. Dr., Institut für Pflanzenbau und Saatgutforschung, Forschungsanstalt für Landwirtschaft, Braunschweig

EBER, W., Institut für Angewandte Botanik der Technischen Universität, Berlin

ELLENBERG, H., Prof. Dr., Systematisch-Geobotanisches Institut der Universität, Göttingen

FUNKE, W., Universitätsdozent Dr., II. Zoologisches Institut der Universität, Göttingen

GEYGER, E., Dr., Systematisch-Geobotanisches Institut der Universität, Göttingen

GNITTKE, J., II. Botanisches Institut der Universität, Gießen

GÖTTSCHE, D., Dipl.-Holzwirt, Institut für Weltforstwirtschaft, Reinbek

HARTMANN, F. K., Prof. Dr., Hann.-Münden

HELLER, H., Dr., Systematisch-Geobotanisches Institut der Universität, Göttingen

KIESE, O., Dipl.-Geogr., Institut für Meteorologie und Klimatologie der Technischen Universität, Hannover

KUBIENA, W. L., Prof. Dr. († 28. 12. 1970), Bundesforschungsanstalt für Forst- und Holzwirtschaft, Reinbek

KUNZE, Chr., Dr., II. Botanisches Institut der Universität, Gießen

LANGE, O. L., Prof. Dr., Botanisches Institut II der Universität, Würzburg

MAYER, R., Dipl.-Geol., Institut für Bodenkunde und Waldernährung der Universität, Göttingen

MEYER, F. H., Prof. Dr., Institut für Landschaftspflege und Naturschutz der Technischen Universität, Hannover

NIESE, G., Dr., Institut für Landwirtschaftliche Mikrobiologie der Universität, Gießen

RUNGE, M., Dr., Systematisch-Geobotanisches Institut der Universität, Göttingen

SATOR, Chr., Institut für Pflanzenbau und Saatgutforschung, Forschungsanstalt für Landwirtschaft, Braunschweig

SCHOBER, R., Prof. Dr., Institut für Forsteinrichtung und Ertragskunde der Universität, Göttingen

SCHULZE, E.-D., Dr., Botanisches Institut II der Universität, Würzburg

SEIBT, G., Dr., Niedersächsische Forstliche Versuchsanstalt, Göttingen

SPEIDEL, B., Prof. Dr., Hessische Lehr- und Forschungsanstalt für Grünlandwirtschaft und Futterbau, Bad Hersfeld

STEUBING, L., Prof. Dr., II. Botanisches Institut der Universität, Gießen

ULRICH, B., Prof. Dr., Institut für Bodenkunde und Waldernährung der Universität, Göttingen

ULRICH, M., Institut für Bodenkunde und Waldernährung der Universität, Göttingen

WEIDEMANN, G., Dr., II. Zoologisches Institut der Universität, Göttingen

WEISS, A., Hessische Lehr- und Forschungsanstalt für Grünlandwirtschaft und Futterbau, Bad Hersfeld

WINTER, K., Niedersächsische Forstliche Versuchsanstalt, Göttingen

# Introductory Survey

H. ELLENBERG

## I. Ecology and the International Biological Program

### 1. The Biological Basis of Productivity

We begin to think about how complicated a motor or other technical system is and how its essential parts interact, only when it no longer functions normally; when its performance declines or suddenly ceases. Our reaction is similar when it comes to the natural systems on which we depend, be it the human body and its environment, a stand of plants in field or garden, or the sea with its swarms of fishes. Only since various kinds of disturbances have begun to appear in these natural systems as a result of contamination, indeed poisoning, of air, water and soil, do we feel alarmed. And we stand stunned in recognition of the fact that we know far too little about these vital systems.

What do we know about the forces, materials and organisms involved in producing the grain we have harvested every year for centuries, in providing good nourishment for the cattle grazing on our meadows, and in enabling the wood to grow, on which we still depend as a raw material despite multitudinous new substitute products? What do we know about the biological balance in our fields, in our streams and rivers, or in the sea? Despite great progress in physiology, molecular biology and biochemistry, despite extensive knowledge about plant diseases and animal pests, despite all the advances made in agriculture and forestry, as well as in fishery biology, we still have no clear understanding of the "biological basis of productivity". This is the reason for the initiation of an "International Biological Program" (IBP) upon this very theme, so that we can try to close the gaps in our knowledge before it is too late.

Most of the plant and animal products which we eat or use industrially originate in biological communities in which not only the organisms of direct interest to us, but also many others, usually invisible, can be of decisive importance. It is about these very relationships that we are so poorly informed today. For this reason, the central emphasis of the IBP is concentrated on organisms and their environment. In fact, it should really have been named the "International Ecological Program", but the planners feared this would not be understood by the general public: so little has "ecology", the science of the relationships of organisms to one another and to their environment, become a generally familiar term. It is beginning to gain wider understanding today as "environmental research".

### 2. Research in Terrestrial Biological Communities

Since the largest gaps in our knowledge exist in the sector of the terrestrial biological communities, the West German national committee for the IBP decided to concentrate all its personnel and funds in this area and to support the studies of freshwater and marine biological communities only within the framework already existing in Germany. In order to give a vigorous stimulus to biological community research, the Deutsche Forschungsgemeinschaft (German Research Association) created the "Central Program: Experimental Ecology" in 1966. The aims and organization of this program have already been reported elsewhere (ELLENBERG, 1967).

Next to that of Belgium, West Germany's was the first comprehensive pilot project to be initiated under the auspices of the IBP. Both are based on guidelines drawn up in Brussels in the fall of 1963 and published in 1964 (see ELLENBERG and OVINGTON, and IBP News No. 2). In the meantime, many other countries have decided to set up similar ecological research programs, among them the U. K., France, the Netherlands, Denmark, Poland, Czechoslovakia, Hungary, Italy, the U.S.S.R., Japan and – for grassland mainly – Canada, and recently also the U.S.A., Brazil, Norway, Sweden, Finland, Yugoslavia, Spain etc.

These newly formed research groups, as well as some still in the process of formation, have asked us to publish our methods and results as soon as possible and in English, so that this information will be generally accessible. Where possible, we have also included interim results in the accompanying joint publication. The final synthesis of our results is not expected to be available until 1973.

Springer-Verlag agreed to publish our manuscript as Volume 2 of their newly created series, "Ecological Studies". We thank them for this, as well as for the patience and consistency with which Springer's English copy-editor, Mrs. B. M. CROOK, corrected our sometimes deficient English.

## II. Structure and Functioning of Ecosystems

### 1. General Scope

The natural interaction of organisms and their environment is called an "ecosystem", a term introduced by TANSLEY (1935). "Biogeocoenosis", a term coined by SUKACHEV at about the same time, is also applicable here. It depicts a more or less complex interaction between biological community (biocoenosis) and habitat (biotope, expressed in the above word by the syllable "geo").

The individual ecosystems vary greatly in size and structure. The entire globe is an ecosystem, the only one which is not influenced by other ecosystems. An island, a forest, a pasture, a decaying tree stump with its mosses and fungi, even a puddle on the path which is only temporarily inhabited, all such natural phenomena deserve to be called ecosystems. Thus, great variations exist not only in magnitude, duration and production, but also in the degree of dependence on other ecosystems.

Apart from a few very divergent attempts, no systematic survey of this multiplicity has yet been made. Such a survey can only be compiled when we know more about ecosystems and when we have more closely examined at least a few of them in respect to their organization, inner relationships, production, and dependence on external conditions.

Such comprehensive investigations in carefully selected project areas are being undertaken by several research groups in the IBP; among these is the West German group. In order to acquire data of a more widely applicable nature, in addition to

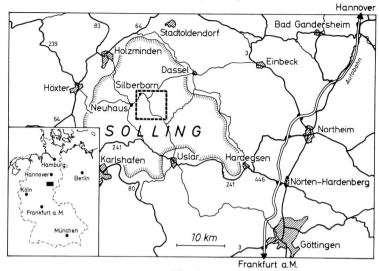

Fig. 1a

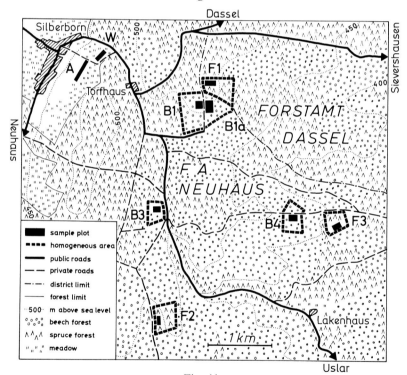

Fig. 1b

Fig. 1. Situation of the research area and of the experimental plots on the Solling plateau, 55 km NW of Göttingen. **B** beech *(Fagus silvatica)* forests, deciduous; B 1 and B 1a old, B 3 younger, B 4 youngest stand. **F** spruce (German „Fichte", *Picea abies*) forest, evergreen; F 2 old, F 1 younger, F 3 youngest stand. **W** meadow (German „Wiese", *Trisetetum flavescentis, Festuca rubra* facies), a mown grassland. **A** arable field (with *Zea mays* and *Lolium multiflorum* resp.)

those valid for the individual ecosystem in question, we chose four adjoining plant formations which, under identical conditions of soil and general climate, form several ecosystems which are significant for the West German landscape:

a) deciduous beech forest (B in Fig. 1),

b) evergreen spruce forest (F),

c) grassland under different fertilizer treatments (W),

d) cultivated field with annual plants whose treatment and fertilizer dosages varied (A).

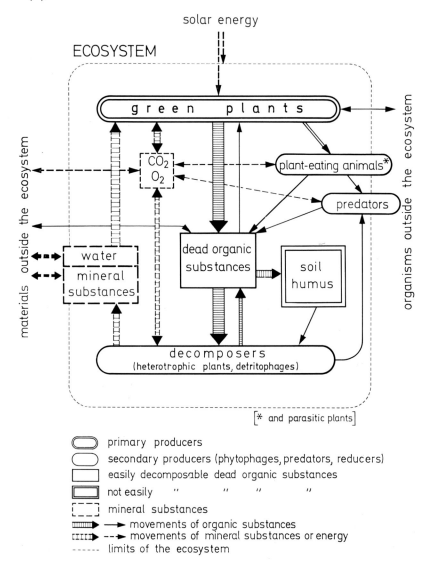

Fig. 2. General scheme of an ecosystem. The individual components and processes may be of varying importance (see section II)

## 2. Primary Producers

What do these ecosystems have in common with those named above and with many others one could name, e.g. semi-deserts, tundras, steppes, savannas and tropical forests? What justifies their being grouped together under one general term? Their principal common denominator is the interaction of the links of one or more food chains. These always begin with the "primary producers", usually green plants which, with the aid of solar energy, produce carbohydrates and ultimately build up other organic substances (see Fig. 2).

## 3. Secondary Producers: Consumers and Predators

The "secondary producers" include plant-eating animals or parasites which feed on plants, e.g. ruminants, amoebas and insects. These "consumers" serve as food for the predators, e.g. carnivorous insects or birds, and these in turn may be eaten by other predators. Man, like many animals, is omnivorous and so fits into such food chains at various places; thus he has increasingly become the superior competitor. Toxic substances can become concentrated in the food chains and become dangerous, even to man.

Although, quantitatively speaking, the predators usually play only a small role, they can be of decisive significance for the stability and production of some ecosystems in that they control the number of plant-eaters. The mechanism of such biological balances has, however, not yet been satisfactorily investigated under natural conditions. Such studies are especially difficult in respect to methodology because some consumers are mobile and often do not remain in a single ecosystem, but live there only temporarily; accordingly, the "plant-eaters" and the "predators" in Fig. 2 are, to some extent, active beyond the borders of the schematically outlined ecosystem. Furthermore, the study of biological balances is rendered difficult by the fact many animals go through successive stages of development in various, often widely separated, biotopes.

## 4. Decomposers

Theoretically, the consumers of living plant material and their enemies are not necessarily an essential component of every ecosystem and are almost completely absent from some, e.g. in the system comprising a cultivated field protected from pests (d). The third group of secondary producers, collectively known as "decomposers", is however indispensable. To this group belong the plants and animals which live on dead organic material (e.g. litter or root remnants), especially fungi and bacteria. Without continuous decomposition and ultimate mineralization of dead organic material, vast quantities of undecomposed remains would pile up in our woods and other plant formations, as well as in stagnant or sluggishly flowing waters. This "natural garbage" would ultimately make it impossible for the plant roots to reach mineral soil and, as is the situation on a raised bog, would force them to get along with less and less nutrient materials. As a consequence, the productivity of most soils, under natural conditions nearly constantly quite high, would fall off considerably.

Primary producers and decomposers are therefore essential components of every independent ecosystem, be it on land or in the water. Where one of these groups is absent, the ecosystem is dependent on other ecosystems, e. g. for the provision of organic material when no green plants are present, or for their removal when decomposers are lacking. Good examples of the first case are the "intestinal flora" of a mammal, or the abundant biological community of a river estuary. The classic example for the second case is the raised bog mentioned earlier whose *Sphagnum* mosses, on the substratum of their compressed but barely decomposed dead parts, grow up higher and higher above the general ground water level.

## 5. Mineral Cycling

Of course, without a certain minimal level of nutrition, the raised bog ecosystem would not be able to exist; the nitrogen, phosphorus, potash, calcium and other substances necessary for the normal growth of green plants, which are collected in the lower layers, are soon no longer accessible to the upward-growing mosses and the vascular plants which accompany them. The biological community of the *Sphagnum* bog acquires these materials from the air, together with the rainwater on which its water supply depends.

With the exception of these and other extreme cases, it is characteristic of all ecosystems that they maintain the circulation of materials, by which the growing plant stand (or plant plankton) is connected with the mineral soil (or open water), via their debris and the corresponding decomposers. The consumers are also included in these cycles, particularly the predators which feed on detritus-eating terrestrial animals. Depending on the local conditions, these cycles can terminate rapidly or continue for many years. Their extension can differ greatly. Here, too, we know very little, especially about the circulation of nitrogen and phosphorus, which are particularly important factors in production. For this reason, special attention is being given to these elements in the investigations reported here.

Man alters these cycles in his favor by removing tree trunks or other components from the ecosystem (e. g. in a and b), or by cutting down nearly the entire plant community above ground (c and d). If he wishes to avoid lowering the productivity, he must replace the materials removed from the soil by adding organic or mineral fertilizer. Fertilization thus normally serves only to maintain or restore the natural fertility, not to increase it. Conditions of soil and climate being the same, only when extremely high dosages of fertilizer are applied does the productivity exceed the level possible in the natural ecosystem. A comparison of research plots a — d demonstrated this fact.

## 6. Water Cycling

Water is an indispensable substance, constantly required in large amounts for the life activities of all organisms, but particularly for photosynthesis in green plants and for the circulation of nutrient materials. Thus, the water supply, along with warmth and food supply, is a most important environmental condition. Because the existence of all ecosystems depends on the water supply, it is essential to include it in the investigations of these systems. Like radiant energy, water is supplied to most systems from outside the system itself. At least in the non-tropical latitudes, circulating air often brings precipitation from afar and carries away local water vapor. In addition, water is fed into and out of some ecosystems either above or below ground level.

Thus, apart from the situation presented by the Earth, or by a broad climatic belt in its entirety, the water cycle is involved only to a very small extent within any particular ecosystem.

## 7. Energy Flow

Food chains and the circulation of water and materials bind the components of every ecosystem closely together. This relationship can also be expressed by the flow of energy which originates as solar radiation, passes through organisms, and finally is discharged into the inorganic environment. Normally, the flow of energy is measured in calories per unit of time and weight, or per unit of time and unit of ground surface area encompassed by the ecosystem. If the energy content of the various plant and animal products is known, the productivity of the individual producers can also be expressed as the amount of dry matter per unit of time and surface area.

Solar energy is primarily utilized through photosynthesis in green plants. Less than one-half the solar energy which reaches the earth is photosynthetically effective, and a large percentage of that amount remains unused where the seasons are too cold or too dry to allow maximal plant activity. During the vegetation period, most plant communities on land and in the water use about 1–5% of the effective light, seldom more. However, when compared with present technical yields, this effectiveness is not as low as it might appear at first glance. No chemist is as yet capable of producing nutrient and raw materials photosynthetically, and the mechanical utilization of solar energy is still very rare.

Thus, it may be seen that ecosystems, and especially the green plants in them, are of immeasurable value to mankind. Furthermore, they provided most of the raw materials for the coal and petroleum from which we still extract the greater part of our technical energy today. Only atomic energy is capable of making our technology independent of the energy-binding capacity of plants.

How given environmental possibilities are utilized by the various ecosystems, what they produce, and the quality of what they produce, will always be of vital consequence for human nutrition. Thus, studies of the *"biological basis of productivity"* do indeed serve *"human welfare"*, as stated in the central theme of the IBP.

# III. Organization of the Integrated Research

## 1. Cooperative Work

The study of one, much less of several ecosystems cannot be accomplished by a single individual. Even when the research is limited to the essential partners and environmental factors, and the circulation of energy, water and the most important elements are to be covered on the basis of only a few major points, professionals are needed who will work in close collaboration with one another. Unlike the situation observable in many other research projects, their work is not connected only by time and location, but must be developed into an effective entity.

Such a synthesis is not easily achieved and can only be successful if all involved are willing, from the outset, to comply with the wishes of the other participants, as well as to exchange data and mutually evaluate the collective results. This close cooperation does not exclude the possibility of each participant identifying his special research interests with the common goals. Indeed, experience shows that such cc-operative research functions only when the individual does not feel he is merely taking orders from a superior planning authority, but feels rather that his own research plans are being supported by the study group as a whole, or at least by some of its members. These prerequisites were fulfilled from the outset in the Solling Project of the German Research Association. Numerous institutes and individuals had declared themselves willing to collaborate on the project.

There were, however, difficulties in finding for each important sector a participant who was willing and able to dedicate himself intensively enough to that particular area. Consequently, we have had to put up with gaps here and there which are also noticeable in this publication. Our cooperation hinders no-one from publishing his own results alone and as early as he wishes. In cases where he depends on the data of others, he must often be patient for a while. It was planned from the outset to undertake major synthesis only after several years of parallel work. Although some of us have been working on the Solling Project since 1965 and nearly all of us since 1967, very little has been published so far.

Furthermore, it was prudent not to be too hasty in publishing data, since variations occur in the environmental conditions and, consequently, in the productivity of the ecosystems, which means that a single year's data cannot be a basis for valid judgement. The entire "Central Program: Experimental Ecology", of which the Solling Project is a part, has been planned to span 5–7 years; the cooperative field work is to be terminated in the fall of 1972.

This book represents our research group's first publication of a more comprehensive nature. Among ourselves, however, we have exchanged and discussed methods, descriptions and data from the very beginning. Some of these descriptions were prepared in English so that they might aid other study groups in the IBP. On the whole, a welcome and much appreciated interchange of ideas was started with colleagues in neighboring countries, especially with those in Belgium (under the direction of DUVIGNEAUD and GALOUX), Denmark (THAMDRUP), Poland (MEDWECKA-KORNAS), Czechoslovakia (JURKO and others) and the United Kingdom (SATCHELL).

## 2. Synthesis of Results

In order to coordinate the work and to exchange information on data and problems, symposia for all participants were held every year in the spring and occasionally in the fall. These symposia usually consisted of 2–3 days filled with reports and discussions. When held in Göttingen, they always ended with discussions on the project site in order to allow collaboration without mutual disturbance.

A special type of symposium was held in April 1970; its central theme was: "Preparation for the Synthesis of all Data". Its program is given here in shortened form, since it summarizes the common goal of all the articles in this book.

Essentially identical habitats in the Solling area comprising
a) beech forests,
b) coniferous stands,
c) variously fertilized grasslands and
d) cultivated fields

are to be compared with one another with the aim of determining characteristic differences and similarities, as well as obtaining more generally applicable data, and working out model ecosystems. The main points of common interest may be grouped as follows:

*Basic data on ecosystems:*
1. Inventory of species and structural descriptions of the communities;
2. Variations in the abundance of important species and sub-species over time;
3. Biomasses and effective surfaces of the main components;
4. Biotope (climate, weather, topography, soil profile, human influences, etc.).

*Magnitude of production by important partners in the biological communities:*
1. Photosynthesis in the various layers of the stand;
2. Dry-matter production of the plant and animal populations;
3. Composition and quality of plant and animal products.

*Exchanges and cycles:*
1. Energy exchange in the ecosystems;
2. The water cycle (over the period of one year and over shorter periods of time);
3. The carbon cycle;
4. The cycling of primarily soil-related elements (N, P, Ca, K, Na, etc.).

*Correlations whose validity is assured:*
1. Life cycles and environmental factors;
2. Magnitudes of production and environmental factors;
3. Magnitudes of production and easily obtained population data.

*Practical conclusions:*
1. Production-limiting factors and possibilities for increasing production;
2. Spheres of validity of the data obtained;
3. Recommended program for making production analyses at minimal expense.

Our common goal was to answer the following questions for each of the five groups of points listed above:

What data are already available?

What results can be expected by the conclusion of the field work, i.e. by the end of 1972?

Where are there strategic gaps and who could fill them?

Where should cooperation be intensified?

What measures must still be taken in respect to cooperative data analysis?

Where is help needed from other IBP study groups?


## IV. The Research Area and the Experimental Plots

### 1. The Solling Region

Before describing the research project and its methods in detail, a short description of our sample areas and their respective, generally accessible, equipment is in order. More complete inventories of the biological communities are to be included in a later publication.

The sample plots are located in an area only a few square kilometers in extent on the plateau of the High Solling, a mountain range of medium altitude about 55 km NW of Göttingen (Fig. 1). Thus, they are centrally located for co-workers from the universities or other institutions in Hamburg, Braunschweig, Hannover, Göttingen, Hann.-Münden, Bad Hersfeld, Giessen, Frankfurt, Würzburg, Freiburg and Berlin, all in West Germany).

The Solling plateau was chosen, however, primarily because it represents one of Germany's largest deciduous forest areas, dominated by acidophilous beech forest

Fig. 3. View into the beech experimental area B 1 with the instrument tower, corner stakes and other equipment. Note that there is hardly any undergrowth

(*Luzulo-Fagetum*, Fig. 3). This is the most abundant naturally occurring plant community in West Germany, and is still the most important forest community in this area today.

In the Solling, as in all of West Germany, the natural deciduous forests are being replaced at an increasing rate by artificially cultivated spruce (*Picea abies*, Fig. 4). Evergreen conifers are favoured by foresters because they are faster-growing than the naturally dominating deciduous trees. They are, however, influencing the water and nutrient cycling, as well as other qualities of the ecosystems, and may result in a lowering of the natural productivity. Since this problem has not been tackled so far, it was included in our integrated research work. In the Solling area there are pure stands of beech and spruce of various age groups, thus making it possible to investigate the chronological development of their productivity on the basis of sample areas in close proximity to one another (see Fig. 1).

Since all these areas belong to the State of Niedersachsen, it was relatively easy to remove them from normal management for scientific purposes. The same is true of the grassland and cultivated fields included in our program.

A further advantage for our cooperative work was the fact that all biological communities in the Solling area are relatively poor in plant and animal species. In the woodland, as in the grassland, very few plants and animals play dominant roles. It is, therefore, possible to determine their quantitative significance in the ecosystem economy.

Fig. 4. Part of the spruce sample area F 1. Here the undergrowth is dominated by *Vaccinium myrtillus* and *Dryopteris carthusiana*

This lack of abundant plant and animal life in all the Solling sample areas is connected with the near constancy of existing climatological and soil conditions. The general climate is characterized by abundant rainfall (approximately 1100 mm annually) and is slightly oceanic in nature, thus presenting favourable conditions for leaching of the soil. Although the sample areas lie only 500 m above sea level, their climate is decidedly montane. Fog occurs frequently at all times of the year, and the winters are long and bring much snow; hoar frost is not rare. The average annual air temperature in the shade at a height of 2 m is only 6.5° C.

The geological substratum of the entire plateau consists of Triassic red sandstone (Buntsandstein) which, due to weathering, is rapidly becoming deficient in calcium. This is covered with loess loam which, though extremely acid, possesses an excellent water-holding capacity; its thickness varies between 50 and 80 cm. This silty loam and the partially clayey sandstone beneath it were more or less transposed and homogenized by solifluction during the last glacial period. Typologically, this soil must be classified as strongly acid brown forest soil (stark saure Braunerde) which, due to local stagnant water, shows occasional characteristics typical of gley soils (for profile description see BENECKE and MAYER, chapter U II).

## 2. The Individual Sample Areas

Fig. 1 shows the locations of the individual sample areas; the descriptions which follow are deliberately sketchy and are meant only to assist the understanding of the articles in this book.

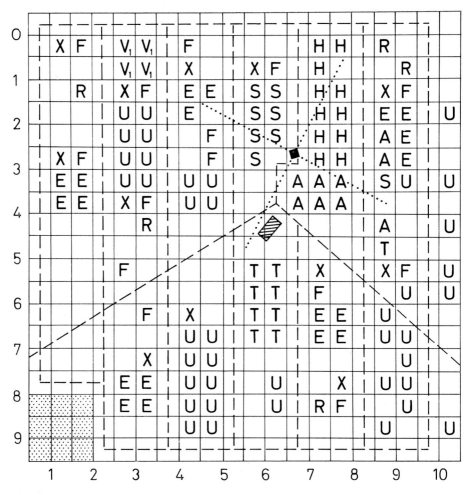

Fig. 5. Map of sample area B 1, showing the position of the measuring tower and of the hut with recording equipment, the footpaths, the 5×5 m grid, and the 15×15 m "typical" area. Sample areas of individual co-workers are marked with letters according to the chapters in this book ($V_1$ = Va.)

Since all the sample areas, and especially the main ones, B 1 and F 1, must be visited innumerable times by many co-workers over the course of the years, it was necessary to establish strict rules from the very beginning. In order to prevent mutual disturbance and to avoid rendering the common research object completely useless, the main forest sample areas have been divided into squares which were marked by permanent corner stakes. Several squares, randomly selected from the entire sample area, were delegated to each co-worker (e. g. see Fig. 5). Lines conspicuously marked by strips of yellow plastic serve as footpaths. Even on the sample areas in the grass-land and cultivated fields, each co-worker must comply with a plan drafted in joint consultation.

Most areas are protected by a sturdy fence against wildlife. Approach roads and the electricity supply are being extended and laboratory facilities have been installed in the shed of the Ranger Station, Torfhaus.

### a) Beech Forest Plots (B1, B3 and 4)

B 1 lies in compartment 51 of the Forest District Neuhaus, about 500 m above sea level.

The surface of the area is nearly horizontal, sloping at maximum 2° S. The *fenced-in-area* of 100 by 100 m size, is level. A 5 by 5 m grid with numbered stakes was superimposed on the 1 ha sample plot to facilitate mapping of individual stems and crown projections and to determine the exact locations of sampling spots. The stakes stand 50 cm above the ground so that they will also be visible during normal snow cover. All the trees were numbered for a mensurational assessment, and in this way they also form an additional aid for orientation. Since the *area surrounding the fenced plot*, in size about 20 ha, shows nearly the same conditions, investigations which are not suited for the fenced plot can also be made outside.

The sample site is relatively uniformly stocked with red beech trees *(Fagus silvatica)*, 23 to 26 m tall and about 120 years old. The trees show still a rather dense closed canopy. Therefore, there is no shrub stratum and the herbaceous layer is only sparsely developed, mosses occur only on the numerous old stumps. Phytosociologi-cally the stand is typical of the montane *Luzula*-beech forest *(Luzulo-Fagetum)*.

The soil is in general normally drained and belongs to the type of acid brown earth. Temporary stagnation of water in some local depressions has led to the for-mation of concretions at 50–70 cm depth at the these places. In such localities one finds a fern-rich undergrowth of the *Luzulo-Fagetum*. This will generally be ignored in the ecological investigations.

Early in November 1966 a lightning-proof *instrument tower* (Fig. 3) was erected for meteorological and ecological measurements. The tower was placed in such a way that it did not disturb the crown canopy. The tower is made almost entirely of alu-minium pipes which were bolted together with various couplings. It is separated into 4 m high platforms and fastened with steel ropes on all four sides (Fig. 5). The tower extends with its upper platform to a height of 28 m. Thus, it reaches about 2 m above the crown canopy. Individual branches reach out as high as the upper railing. Currently the tower is extended upwards to a height of 36 m by a single aluminium pipe. This is removable for the fixing of recording devices and for checking. Smaller, removable platforms in the form of metal grids are fixed above each of the lower main platforms.

Adjacent to the tower stands a lockable shed $7.5 \times 2.5$ m $\times 2.5$ m high, which houses electrical recording devices (see Fig. 5). Electric power supply comes through a 1300 m long ground-cable from the Forest Ranger Station Torfhaus.

For incorporation of the sample plot into the general net of meteorological stations, a standard instrument shelter was established.

Disturbance of the undergrowth vegetation and soil by frequent trampling is avoided by the use of fixed pathways. These can be replaced, where necessary, by narrow, temporary board-walks. The requirements of the individual investigators will be reconsidered at the regular annual symposia. A smaller subplot $15 \times 15$ m in size, designated as "typicum" of the investigated community, was established as a fenced enclosure within the $100 \times 100$ m plot. This area is "off-limits" to everyone.

Near B 1, in the adjacent comp. 50, a 2 ha area of the same beech stand has been fenced in (B 1a, see Fig. 1) and is off limits to everybody except some zoologists. Soon after the integrated field work was begun, this additional research area proved to be necessary because animals are extremely sensitive to disturbances due to human activities.

B 3 and B 4 are situated in comp. 91 and 27 resp. of the Forest District Neuhaus. As is shown in Table D 1, these beech stands are younger (about 80 and 60 years old). Like all other sample plots in the High Solling, they have climate and soil conditions very similar to those in B 1.

### b) Spruce Forest Plots (F 1—3)

F 1 is not far from B 1 on nearly level ground in comp. 28 of the Forest District Dassel (see Fig. 4). The *Picea abies* stand was planted on grassy heathland. When young, it was relatively little damaged by deer, and is now 85 years old. So far, only a standard meteorological shelter has been established as a permanent installation on this site; but now, an instrument tower (of the same construction as that in B 1) is being erected in autumn 1970.

F 2 and F 3 are spruce stands about 115 and 40 years old. They are situated in the compartments 83 and 25 of the Forest District Neuhaus (see Fig. 1 and Table D 1).

### c) Grassland Areas (W 1 and 2)

W 1. The main sample site belongs, like the other treeless areas and the forested areas, to the Forest District Neuhaus/Solling and has so far been used relatively extensively by leaseholders. The area belongs to comp. 163 (subcomp. 19a), and lies about 475 m above sea level. In contrast to B 1 the area dips towards NW with a slope of about 3°. Unfortunately, no other suitable grassland area could be found that had exactly the same exposure as B 1.

The present stand can be phytosociologically classified as a *Festuca rubra* facies of the yellow oat meadow (*Trisetetum flavescentis*). The rather low yields at present can probably be much increased through fertilization (see W 2 and chapter I).

The soil profile is rather similar to that beneath the forested sample sites, and can also be classified as acid brown earth. The crumb-structured surface horizon is remarkably uniformly humous to a depth of 15 cm. This is ascribed to the fact that the area was earlier used as agricultural crop field, a rotation that was customary in the Solling some time ago. However, the area has now been under grass for a long time.

A 10 m high instrument tower is installed for the measurement of microclimate in a similar way to the one in the forest stand B 1. The tower can easily be mounted with sensing antennas. The same type of shed as on plot B 1 was erected for the housing of electric recording devices. It stands about 40 m away from the instrument tower. Electric power is tapped from the standard power line which goes across the fenced plot to the Forest Ranger Station Torfhaus. A standard meteorological shelter is placed on the sample plot as in B 1.

W 2. A fertilization experiment was initiated in the autumn of 1966, which is contained within the fenced plot W 1. The layout is described in chapter I. The 36 treated squares (5 × 5m) are marked by coloured stakes. The fertilization experiment has been extended with the same treatments into the area outside the fence. This area can be used for plant and soil sampling without disturbing the actual experiment subplots.

The experiment was purposely designed for only three treatments with increasing amounts of fertilizer and 6 replicates each. These are sufficient for evaluating the anticipated productivity increases. The same experimental design has been used successfully in many grassland experiments for several years.

The area of the grassland plot (W 1), which was not subjected to the fertilization experiment, or unaffected by instrument installation, will be managed by the lease-holder in the usual way, i.e. it will normally be mown twice a year.

### d) Arable Fields (A1 and 2)

Two crop fields (sections 88 and 89 of compartment 163, Neuhaus) are close to W 1, where they occur under similar climatic and edaphic conditions. They were managed till 1967 in the way customary for this locality by a leaseholder (oat, potatoes and rye alternating). As research plots, they were cultivated with annual grasses *(Zea mays* and *Lolium multiflorum)* and subjected to intensive fertilization and mowing, aiming at maximum yields.

At the ends of the sample plots, weeds were allowed to develop under the same fertilizing conditions, mainly to study the leaf area index (see chapter J).

# References

ELLENBERG, H.: International Biological Programme. Contributions of the Federal Republic of Germany. Bad Godesberg: German Research Association 1967.
— OVINGTON, J. D.: International Biological Programme of the International Council of Scientific Unions. Project A I: Productivity of Terrestrial Communities — Ecology. Ber. Geobot. Inst. ETH, Stiftg. Rübel, Zürich, **35**, 29—40 (1964).
*IBP News.* Ed. by IBP Central Office 1964—1969 (20 numbers).

# A. Measurement of $CO_2$ Gas-Exchange and Transpiration in the Beech (Fagus silvatica L.)

O. L. LANGE and E.-D. SCHULZE

## I. Introduction

The elementary process in all primary production is the photosynthetic assimilation of $CO_2$ by the green plant. The synthesized assimilates of this process are the basis of anabolic and catabolic metabolism of all organisms living in the biotope. The quantity of products photosynthesized is determined by the specific activity of the plant species and depends in a complex manner on the external conditions of the habitat as well as on endogenous factors. The main producer of the *Luzulo-Fagetum* in the Solling (experimental area B 1) is the beech (*Fagus silvatica* L.). To make a quantitative determination of the influx of organic material into the ecosystem due to the productivity of the beech, and to resolve the fine structure of this productivity, the $CO_2$ gas-exchange of leaves and twigs of a 26 m tall experimental tree was continuously recorded over an entire year in both the sunlit and the shaded crown; transpiration was simultaneously recorded. The measurements were made under natural conditions as nearly as possible in order to study the natural reactions of the beech and to correlate them with the dominant microclimatic conditions, which were recorded at the same time. In addition, the gas-exchange was checked under experimentally controlled and systematically varied conditions in the same tree in order to analyze the effect of specific climatic factors and to make a causally based interpretation of primary production.

## II. Measuring Techniques

### 1. Measurement of $CO_2$ Gas-Exchange and Transpiration

The gas-exchange measurements were made with a mobile field laboratory (cp. LANGE, KOCH, and SCHULZE, 1969; KOCH, LANGE, and SCHULZE, 1970) equipped with an infrared gas analyzer, three climatized Siemens gas-exchange chambers and the necessary recording instruments (block diagram of the whole set-up: Fig. 2).

---

Fig. 1a. The installation of two gas-exchange chambers in the sunlit crown of the beech during summer 1968. At each chamber the radial fan can be seen. The tubing runs from the chamber to a box with the humidity sensors. The outside leaf temperature sensor can be seen on the far right

Fig. 1b. Scheme of the gas-exchange chamber. 1 Peltier block; 2 internal radiator; 3 external radiator; 4 connections for additional water cooling; 5 radial fan; 6 cover plate; 7 plexiglass housing; 8 air inlet and outlet for $CO_2$ measuring air and humidity control; 9 cable inlet for thermocouples, lightcells etc.; 10 inlet for plant twig etc.; 11 inside temperature sensor Pt 100

Fig. 1a

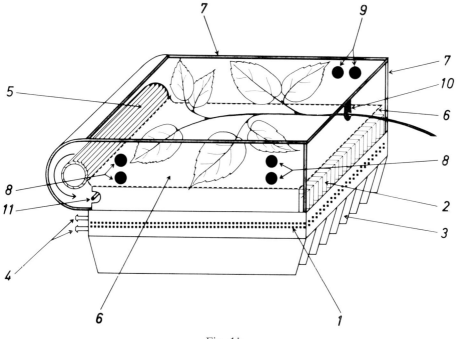

Fig. 1b

For CO$_2$ and H$_2$O gas-exchange measurements during the vegetation period of 1968, twigs with 10—20 leaves still in their natural position were enclosed for periods of 4—5 days (spring and autumn) and 8—10 days (summer) in the gas-exchange chambers (Fig. 1a, b; Fig. 2: GWK, cp. Koch and Walz, 1967; Koch, Klein, and Walz, 1968). To facilitate this, the experimental tree was pulled a little towards the measuring tower and secured so that the branches of both the sunlit crown (26 m high) and the shaded crown (17 m high) could be reached from the platforms where the gas-exchange chambers were mounted.

### a) Plant Chamber and Transpiration Measurement

For the CO$_2$ analysis, a pump (Fig. 2: P$_1$) draws the measuring air through the cuvette to the infrared gas analyzer (URAS). The volume of the chamber made of

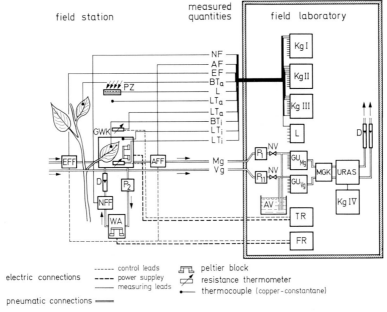

Fig. 2. The layout of the measuring system: *AFF* output humidity sensor, *AF* output humidity, *AV* shutoff valve, *BTa*, *BTi*, leaf temperature outside, inside the chamber, *D* flow meter, *EFF* input humidity sensor, *EF* input humidity, *FR* humidity controller, *GU* gas changeover switch, *GWK* gas-exchange chamber, *Kg* compensation point recorder, *L* light intensity, *LTa, i*, air temperature outside, inside the chamber, *MGK* measuring gas cooler, *Mg* measuring air, *NFF* humidity sensor in the circulating air stream, *NF* humidity in the circulating air stream, *NV* needle valve, *P* pump, *PZ* photocell, *TE* thermocouple, *TR* temperature controller, *Vg* reference air, *WA* water vapour trap (after Koch, Lange, and Schulze, 1970)

perspex (Fig. 1b) is 7 l. The effective space of 280×200×80 mm is separated at the bottom by a metal plate from the Peltier radiator which controls the temperature in the chamber. A radial fan provides a maximum air circulation of 2 m/sec. For the measurements, the air exchange in the cuvette is set according to the local average wind speed at $1-2$ m/sec in the sunlit crown and $0.5-1$ m/sec in the shaded part of the crown. Air temperature in the cuvette is determined either by a "follow-up control" (inside temperature instantly following outside temperature) or a "constant-value control" (inside temperature constant, independent of external conditions). For follow-up control, a representative air temperature is measured with a platinum resistance thermometer at a place outside the chamber not influenced by the instrumentation. This outside temperature is compared with the temperature inside the cuvette (Pt 100, LTa, LTi) by an electronic controller (TR). If there is a temperature difference, the cooling or heating power of the Peltier blocks is adjusted to bring the inside and outside air temperatures to the same value within $^1/_{10}$ of a degree centigrade with virtually no time lag. For constant-value control, the outside temperature sensor is replaced by an adjustable resistance in the electronic controller. This keeps the inside air temperature constant; over a certain range it is independent of external conditions and proportional to the set value of the resistance. The heating power is sufficient to maintain a constant overtemperature, which is 30 degrees higher than the outside temperature, without additional external thermal radiation. The cooling power achieves a temperature of 8 degrees below the outside temperature with full solar radiation.

Humidity control is achieved by compensation of the water vapour released by the plant, making transpiration measurement possible. The amount of $H_2O$ released by mesophytic plants is about 100 times greater than $CO_2$ consumption in the same period. Because of this, the flow rate of the air towards the URAS, which has a differential range of $\pm 75$ ppm, is not enough to flush all the transpired water out of the cuvette (TRANQUILLINI, 1964). Therefore a pump ($P_2$) draws a second constant air stream out of the chamber and pumps it through a Peltier cooled water vapour trap (WA) and back into the chamber. The external humidity is measured with a lithium chloride sensor in the air stream going into the chamber (incoming air humidity sensor EFF) and the internal humidity is measured at the exit towards the URAS (outgoing humidity sensor AFF). As soon as the internal humidity rises by transpiration, the differential signal of the humidity sensors is fed into an electronic humidity controller (FR). It adjusts the dew point of the water vapour trap in the circulating air stream to condense all the transpired water, thus equalizing the humidity difference. Sudden changes in the external dew point of $\pm 6°$ C, which are rare even during thunderstorms, are followed by the control with a time lag of $10-20$ min. Gradual changes in the external dew point and transpiration rate are adjusted rapidly. A constant value control is made possible by passing the incoming air through a second water vapour trap having a constant dew point. However, the dew point in the cuvette cannot be raised above the external air temperature without causing condensation in pumps and tubing.

The transpiration (Tr) can be calculated from the flow rate (DS) and the difference in humidity between the air in the chamber (AF) and the dried air in the circulating airflow after it passes the vapour trap (NF): $Tr = (AF\text{-}NF) \times Ds$. The error in the transpiration measurement was calculated by KOCH, LANGE, and SCHULZE (1970). For *Fagus* having an average transpiration of $1000-2000$ mg $H_2O$/gdw · h, the error is about $\pm 10\%$ to $\pm 16\%$.

2*

## b) $CO_2$ Gas-Exchange

Heated tubes run from the gas-exchange chambers in sunlit and shaded crowns to the field laboratory on the ground near the tower. In this laboratory an infrared gas analyzer (URAS I, Hartmann u. Braun, Frankfurt/M.) measures the $CO_2$ uptake and $CO_2$ release of the enclosed twigs in an open system as the differential between the measuring air from the chambers and the external air (BOURDEAU and WOODWELL, 1965). Measuring and reference air are drawn from the same location, and therefore have the same basic $CO_2$ concentration. The absolute $CO_2$ concentration of the air varies in the summer between 340 ppm at night and 312 ppm in the daytime. The air is drawn in at a constant flow rate by electromagnetic pumps (Reciprotor, Edward, Copenhagen), which have continuous adjustment by changing the constant voltage by means of rheostats. The fine adjustment in the flow rate is made with precision needle valves. In the region of 330 ppm $CO_2$, photosynthesis increases with the basic $CO_2$ concentration. Therefore the flow rate was adjusted so that the drop in $CO_2$ concentration of the air in the cuvette due to photosynthesis is not more than $10-15\%$. This limits the error of the measurement. The continuous measuring and reference air flows of the different test stations were connected with magnetic valves automatically in successive pairs to the differential URAS within a 1 min cycle. Because of the sensitivity of the gas analyzer to water vapour as well as $CO_2$, the air was dried to a constant dew point of $1°$ C with a Peltier cooled gas cooler (Siemens). The flow rate is measured at the exit of the URAS (Rotameter, Rota, Oeflingen). A compensation point recorder (Polycomp I, Hartmann u. Braun) records the $CO_2$ values on a 210 mm scale having a differential range of $\pm 75$ ppm. Differences of 0.3 ppm can be evaluated.

The total error of the $CO_2$ measurement is due to the errors caused by the "noise" of the different instruments, the calibration error, the error caused by changes in temperature and pressure, the error caused by changing flow rates from changing pump pressure. Furthermore, the sensitivity of the analyzer depends on the basic $CO_2$ concentration in the reference air (KOCH, LANGE, and SCHULZE, 1970). An additional error is made by the determination of the reference values (dry weight etc.). These errors add up to $\pm 7\%$ to $\pm 14\%$ of the result having an average deviation of 40—20 ppm at the URAS. With equal amounts of plant material in the chamber, the reproducibility of the measurements and the relative comparison between the different test points is even better. The URAS was calibrated every 10 days with two calibration gases whose differential was determined by Hartmann u. Braun Inc. to be 75 ppm $\pm 2$ ppm. The calibration was checked in between more frequently.

## c) Discussion of the Gas-Exchange Measurements

Since the first field measurements of photosynthesis, there has been much discussion as to how far natural climatic conditions can be attained within a cuvette. Various approaches have been tried to meet this problem of "cuvette climate" (BOSIAN, 1964; TRANQUILLINI, 1957; KOCH, 1957; LANGE, 1962; ECKARDT, 1966). With the construction of the Siemens gas-exchange chamber it became possible to control air temperature and air humidity in a closed volume so quickly that they could follow exactly the natural conditions.

In regard to these two factors the plant has almost the same conditions as outside. However the cuvette still affects other important factors, namely convection and the energy balance of the leaves. In the chamber the twigs are exposed to a relatively slight but constant wind, which is necessary for the temperature control. At this wind speed, after breaking the boundary layer of the leaf, the air exchange has little influence on the $CO_2$ uptake. The energy balance of the enclosed leaves is changed mainly through the plexiglass top and through the Peltier system on the cuvette bottom. This, however, changes the amount of photosynthetically active radiation only slightly. At high solar radiation, the leaves in the cuvette have a more constant overtemperature than outside, where the rapidly changing convective conditions lead to a frequently changing leaf temperature. However, the average leaf temperatures inside and outside the cuvette agree within $2-3°$ C. The effect of the changed energy balance and the simultaneous change in convection conditions on the evaporation rate in the chamber was determined by KOCH, LANGE, and SCHULZE (1970). It was found that with high solar radiation the mean evaporation rate inside and outside the cuvette was almost the same. Therefore, in spite of the still unavoidable difference between the chamber climate and the natural field conditions, the gas exchange of the enclosed plant parts can be regarded as a good approximation of the activity under natural conditions. It was also confirmed that the beech twigs, even after 14 days of enclosure, did not show any changes in their photosynthetic activity and that their development was just as rapid as that of the other leaves of the same tree. The twigs were changed more frequently during spring and autumn only in order to avoid a significant change in reference values. The recorded values of $CO_2$ gas-exchange and transpiration were related to dry weight, leaf area, and chlorophyll content of the enclosed leaves. In relating photosynthetic productivity to $CO_2$ balance, its relation to the dry weight of the assimilating leaves seems to be the most meaningful. Calculations on the energy balance or calculations on the energy transfer by transpiration give preference to leaf area as reference. The efficiency of the photosynthetic apparatus is expressed most clearly in relation to chlorophyll content.

## 2. Measurement of Leaf Temperature and Meteorological Data

### a) Air Temperature

Parallel to $CO_2$ gas-exchange and transpiration at each test point, leaf temperature and other important meteorological data were measured inside and outside the cuvette. They were recorded continuously on a 12-channel 6-range compensation point recorder. In addition the observation of a number of check measurements was necessary for perfect operation of the climate control. For example, it is necessary to measure the temperature of the radiator in the chamber to check condensation. The air temperature was measured with Pt 100 resistance thermometers, as used for the temperature control; air temperature was also measured thermoelectrically with shaded needle-shaped thermocouples.

### b) Air Humidity

The air humidity (dew point temperature) was measured inside as well as outside the chamber with lithium chloride sensors which at the same time are used for humidity control. This dew point measurement depends on the following principle: At a certain overtemperature hygroscopic LiCl has the same vapour pressure as the surrounding air. The wet LiCl solution, which is conductive to electric current, is heated by two heating wires until it dries. At this point it is no longer conductive and automatically shuts off the heating current. The overtemperature at an equilibrium of wet and dry LiCl is proportional to the dew point temperature of the surrounding air and can be measured (LÜCK, 1964).

The advantage of the method is that the overtemperature can be very accurately measured with Pt 100 resistance thermometers (same type as used for air temperature measurement). This humidity sensor can also be used over a wide range (5—95% relative humidity). The sensors were impregnated with fresh LiCl solution every 10—14 days and their signal was calibrated by comparison with a ventilated psychrometer.

### c) Absolute Light Intensity

Light intensity was measured with selenium photocells with platinum-opal filters 1:100 (Lange, Berlin). The photocells are individually calibrated for a 6-channel point recorder which has two ranges with the sensitivity of 0—12 klx and 0—120 klx. At the end of June on a day with high solar radiation 10 klx is equivalent to 0.14 cal/cm² · min (measurement of the radiation with a Kipp solarimeter). The signals of the different cells deviate by about 5% from one to another.

### d) Leaf Temperature

At high solar radiation the leaf temperature can differ considerably from the air temperature. For an adequate evaluation of the $CO_2$ and $H_2O$ gas-exchange, an exact measurement of this leaf temperature is of special importance. Its measurement was made thermoelectrically with the "leaf clamp" described by LANGE (1965) with an improved thermocouple (Fig. 3). Two wires respectively of copper and constantan

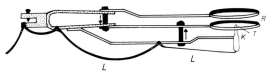

Fig. 3. The leaf clamp for measuring leaf temperature (cp. LANGE, 1965). *L* electrical leads, *K* blade of synthetic resin, *T* thermocouple, *R* clamp

having a diameter of 0.05 mm were soldered together with silver at one end, just at the tip. For mechanical strengthening, the other ends were soldered after about $^1/_2$ cm to 0.3 mm diameter wires of the same metal. The thermocouple together with its leads was cast in a synthetic resin (Araldit, Ciba) so that only the solder point and the first 2—3 mm of thin wire ran along the edge of a very thin blade of resin (Fig. 4). All the other leads were enclosed by Araldit, the heat insulation of which was improved by the addition of some styropor powder. This thermocouple lamella is held by the leaf clamp with its edge pressed against the lower epidermis of the leaf. The temperature-sensitive solder point and the first 2—3 mm of the leads have, on one side of the wire, a good heat transfer to the leaf; on the other side these wires, as well as all the other leads, are heat-insulated against the surrounding air. The error caused by heat conduction (EGGERT, 1946; JARVIS and KAGARISE, 1962) is very small with such thin wires and such insulation. The thermocouple takes the temperature of the leaf tissue to a good approximation. Being measured on the lower side of the leaf, it requires no compensation for the error caused by radiation (RASCHKE, 1954). It is assumed here that within the thin mesophytic leaf of a beech there are virtually no large temperature gradients between the upper and the lower surface of the leaf. The reference solder point was kept constant at $0°$ C by a Peltier cooled metal block

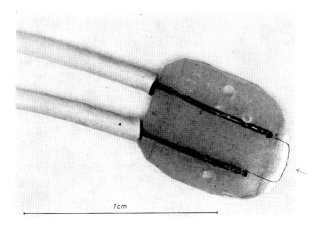

Fig. 4. Copper constantan thermocouple embedded in a blade of synthetic resin. The arrow indicates the soldering point

(Thermonullstelle, Siemens). During the measurements at each test station the leaf temperature was continuously recorded (referred to a point between the leaf veins in the middle of the leaf) inside the chamber as well as outside under natural conditions.

### e) Water Vapour Gradient and Total Diffusion Resistance

From the measurements of air and leaf temperatures and air humidity, it is possible to calculate the water vapour gradient from the leaf to the surrounding air as the difference between the water vapour saturation at leaf temperature and the humidity of the surrounding air. Dividing this water vapour gradient by the corresponding transpiration gives the total diffusion resistance of the leaf to the diffusion of water vapour (BROWN and ESCOMBE, 1900; LANGE, KOCH, and SCHULZE, 1969). This total diffusion resistance comprises the internal resistance of the leaf (stomatal, cuticular and mesophyll resistance) and the external resistance of the aerodynamic boundary layer. As long as the convective conditions inside the chamber are kept constant by running the radial fan at the same speed, the external resistance of the boundary layer is approximately constant with respect to the same leaves. Changes in the total resistance can only be caused by changes in the internal resistance of the leaf. A change in the diffusion resistance can therefore be caused either by a change in the stomatal resistance or by a change in the mesophyll resistance (e. g. WHITEMAN and KOLLER, 1967). According to FISCHER (1968), however, the mesophyll resistance of mesophytic leaves is small, and experiments done on stomata models (LEE, 1967) also make it clear that rapid changes in resistance are caused primarily by movements of the stomata. The gas-exchange of beech leaves is controlled by stomata which are only on the lower side of the leaf. This simplifies the transfer of the $H_2O$ diffusion resistance to the conditions of $CO_2$ diffusion (BERTSCH and DOMES, 1969).

## III. Analysis of the Measurements

In each chamber $CO_2$ analysis was done once every 6 min, and the humidity values for the transpiration determination and the temperature measurements were recorded at each test point every 48 sec. The light intensity was recorded every 2 min at each station. For the evaluation of the data, each $CO_2$ measurement was read off the chart. The temperature, light, and humidity values were averaged visually over 10-min periods. The water vapour gradient and the diffusion resistance were also calculated for 10-min periods. The measurements allow the plotting of the daily natural courses of net photosynthesis and dark respiration of beech leaves and their transpiration in the sunlit and shaded crown respectively during the whole vegetation period (March 1968–December 1968). Fig. 5 gives an example of the daily course of the gas-exchange related to dry weight, leaf surface area and chlorophyll content plotted together with the climatic data. For the montane climate of the Solling, it is a relatively warm, dry July day. The net photosynthesis of the sunlit leaves shows in its daily course almost a two-peaked curve. Transpiration follows the light conditions as well as the water vapour gradient. Compared to more moist conditions, the diffusion resistance is raised during the middle part of the day and transpiration is therefore reduced. The shaded leaves receive only about $1/10$ of the light intensity of the sunlit leaves. Their net assimilation closely follows the light conditions and every patch of light is used for a further increase of the $CO_2$ uptake. The diffusion resistance is very low and allows maximal $H_2O$ as well as $CO_2$ diffusion. Related to dry weight, the net assimilation appears to be almost the same in shade and sun leaves. In relation to leaf surface area and chlorophyll content, the shade leaves assimilate about half as much as the sun leaves, although having only $1/10$ of their light intensity. Transpiration of the shade leaves, related to dry weight as well as to surface area, is lower than that of the sun leaves according to the differing water vapour gradient.

Fig. 5. The daily course of the $CO_2$ and $H_2O$ gas-exchange and the climatic factors for sun (So) and shade (Sch) leaves of beech during July 13, 1968. The diagrams of the figure demonstrate from the top to the bottom: 1 light (klx); 2 net assimilation related to dry weight (mgCO$_2$/gdw · h); 3 related to surface area (mgCO$_2$/dm² · h); 4 related to chlorophyll content (mgCO$_2$/mgChl · h); 5 total diffusion resistance (h/dm · 10⁻²); 6 transpiration (Tr) related to dry weight (mgH$_2$O/gdw · h) and water vapour gradient (WG) (mgH$_2$O/l); 7 transpiration related to surface area (mgH$_2$O/dm² · h); 8 air temperature in the chamber (LTi) and leaf temperature (BTi) (°C); 9 relative humidity (%rF); solid lines: sun leaves, dashed lines: shade leaves (after SCHULZE, 1970)

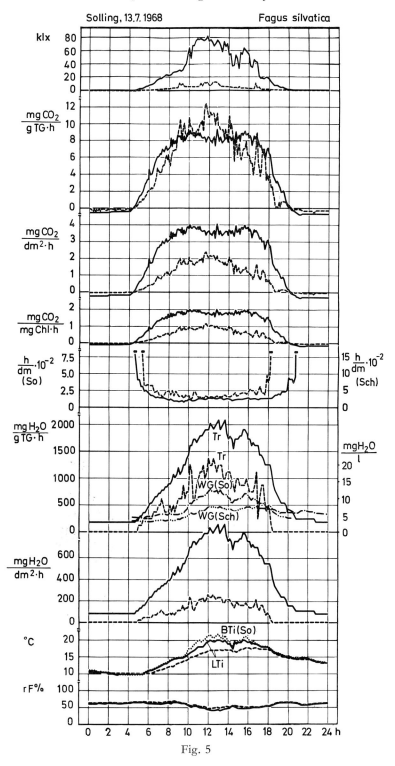

Fig. 5

To analyze the light, temperature, and humidity dependence of the $CO_2$ and $H_2O$ gas-exchange, it is necessary to separate the values of the natural climate recorded over a certain period according to the temperature, light and humidity conditions.

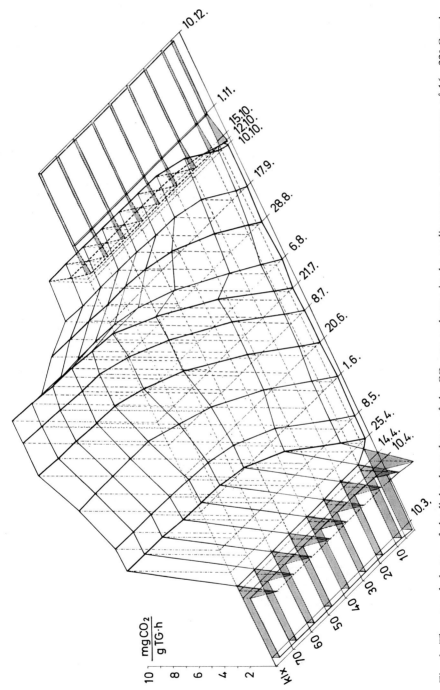

Fig. 6. The annual change of the light dependency of the $CO_2$ gas-exchange in the sunlit crown at temperatures of 16—22° C and a relative humidity of 80—100%. x axis: date; y axis: light intensity; z axis: $CO_2$ gas-exchange. The shaded dark area describes the $CO_2$ release by the respiration of the buds (after Schulze, 1970)

The leaf temperature range was divided into classes of 2° C. The range of light intensity was divided into classes of 10 klx for the sunlit canopy, and 1 klx for the shaded canopy. The water vapour gradient between the leaf and the surrounding air was divided as follows: 0—2, 2—4, 4—10, 10—20 mg $H_2O$/l; a saturation deficit in the air of 20 mg $H_2O$/l is equivalent to 30% relative humidity at 30° C. For the ecological evaluation of the effect of specific climatic conditions, the relative air humidity was taken as basis and arranged in 20% classes denoted by the corresponding class average of 90%, 70%, 50% and 30%. For the different stages of development throughout the year, each 10 min value within 6—10 day periods of net assimilation was taken from the recording chart and put into the appropriate light, temperature, and humidity class together with its corresponding transpiration measurement and calculated leaf resistance. The basis for further evaluations is the mean of at least 2, but usually 10—20, of such 10-min values. For example, this enables the typical changing light dependence of the net assimilation of beech in its sun-exposed canopy to be evaluated for the year at optimal temperature and humidity conditions (Fig. 6). The diagram shows the changing level of light saturation and gives an idea of the differential efficiency of photosynthesis. Calculations of this kind make it possible to identify the effect of the various climatic factors on $CO_2$ assimilation, to evaluate actual productivity and to determine how it changes under differing climatic conditions. Thus estimates can be made of the extent to which the theoretically possible $CO_2$ uptake of the beech is limited and diminished by the actual climatic conditions at the different stages of the vegetation period. The daily balance of net assimilation and night respiratory losses is used to calculate the daily photosynthetic net gain of the beech leaves which is summed to give the annual balance. Using the biomass of the leaves, it is then possible to calculate the approximate net gain of the experimental tree or of the total forest stand. This assimilatory net gain can be used by the next trophic levels within the ecosystem to enlarge their biomass and for their catabolic metabolism.

The investigations in the Solling were carried out by E. D. SCHULZE. The results are published in detail in another place (SCHULZE, 1970).

# References

BERTSCH, A., DOMES, W.: $CO_2$- Gaswechsel amphistomatischer Blätter. I. Der Einfluß unterschiedlicher Stomataverteilung der beiden Blattepidermen auf den $CO_2$-Transport. Planta **85**, 183—193 (1969).

BOSIAN, G.: Der Peltiereffekt im Einsatz zur Küvettenklimaregulierung. Ber. Deut. Botan. Ges. **77**, 22—23 (1964).

BOURDEAU, P. F., WOODWELL, G. M.: Measurements of plant carbon dioxide exchange by infrared absorption under controlled conditions and in the field. Methodology of plant eco-physiology, Proceedings of the Montpellier Symposium, Unesco 1965, pp. 283—289.

BROWN, H., ESCOMBE, F.: Static diffusion of gases and liquids in relation to the assimilation of carbon and translocation in plants. Phil. Trans. Roy. Soc. London B **193**, 223—291 (1900).

ECKARDT, F. E.: Le principe de la soufflerie climatisée, appliqué a l'étude des éxchanges gaseux de la couverture végétale. Oecol. Plant. **1**, 369—400 (1966).

EGGERT, R.: The construction and installation of thermocouples for biological research. J. Agric. Res. **72**, 341—354 (1946).

Fischer, R. A.: Resistance to water loss in the mesophyll of leek *(Allium porrum)*. J. Exp. Botany **19**, 135—145 (1968).

Jarvis, N. L., Kagarise, R. E.: Determination of the surface temperature of water during evaporation studies. A comparison of thermistor with infrared radiometer measurements. J. Colloid Sci. **17**, 501—511 (1962).

Koch, W.: Der Tagesgang der „Produktivität der Transpiration". Planta **48**, 418—452 (1957).

— Walz, H.: Kleinklimaanlage zur Messung des pflanzlichen Gaswechsels. Naturwissenschaften **54**, 321—322 (1967).

— Klein, E., Walz, H.: Neuartige Gaswechsel-Meßanlage für Pflanzen im Laboratorium und Freiland. Siemens Z. **42**, 392—404 (1968).

— Lange, O. L., Schulze, E. D.: Ecophysiological methods for measuring $CO_2$ gas exchange and transpiration in the field — a mobile laboratory for environmental research. (In preparation) 1970.

Lange, O. L.: Eine „Klapp-Küvette" zur $CO_2$-Gaswechselregistrierung an Blättern von Freilandpflanzen mit dem URAS. Ber. Deut. Botan. Ges. **75**, 41—50 (1962).

— Leaf temperatures and methods of measurement. In: Methodology of plant eco-physiology, Proceedings of the Montpellier Symposium, (F. Eckardt, Ed.), Unesco 1965, pp. 203—209.

— Koch, W., Schulze, E. D.: $CO_2$-Gaswechsel und Wasserhaushalt von Pflanzen in der Negev-Wüste am Ende der Trockenzeit. Ber. Deut. Botan. Ges. **82**, 39—61 (1969).

Lee, R.: The hydrologic importance of transpiration control by stomata. Water Resources Res. **3**, 737—752 (1967).

Lück, W.: Feuchtigkeit: Grundlagen, Messen, Regeln. München-Wien: R. Oldenbourg 1964.

Raschke, K.: Die Kompensation des Strahlungsfehlers thermoelektrischer Meßfühler. Archiv f. Meteorologie, Geophysik u. Bioklimatol. **5 B**, 447—455 (1954).

Schulze, E. D.: Der $CO_2$-Gaswechsel der Buche *(Fagus silvatica* L.) in Abhängigkeit von den Klimafaktoren im Freiland. Flora **159**, 177—232 (1970).

Tranquillini, R.: Standortklima, Wasserbilanz und $CO_2$-Gaswechsel junger Zirben *(Pinus cembra* L.) an der alpinen Waldgrenze. Planta **49**, 612—661 (1957).

— Blattemperatur, Evaporation und Photosynthese bei verschiedener Durchströmung der Assimilationsküvette. Mit einem Beitrag zur Kenntnis der Verdunstung in 2000 m Seehöhe. Ber. Deut. Botan. Ges. **77**, 204—218 (1964).

Whiteman, P. C., Koller, D.: Interactions of carbon dioxide concentration, light intensity and temperature on plant resistance to water vapour and carbon dioxide diffusion. New Phytol. **66**, 463—473 (1967).

# B. Estimation of Photosynthetically Active Leaf Area in Forests

H. Heller

## I. Methods for Beech Leaves

Numerous methods have been developed for the determination of leaf area
Marshall (1968) has reviewed and classified the more important ones.

During our studies in the Solling forests, we have used three methods for beech
leaves *(Fagus silvatica)*:

1. measurement with a planimeter,
2. comparison with standards (leaf images),
3. cutting of leaf discs.

The first method was used only for special investigations with small samples;
in particular it was used to determine the annual development of leaf characteristics
at different crown heights.

To estimate the Leaf Area Index (L.A.I., area of one side) one needs a method
which allows rapid determination of the area of many leaves. The second method
was recommended for determining L.A.I. by Ellenberg after satisfactory experience
with it in Swiss forests (unpublished data).

A series of leaf images of known area is selected from the material measured by planimeter, and a standard is made for leaves from 4 to 42 cm² (Fig. 1). Leaves of unknown
area are compared with the scale and classified according to their resemblance to these shapes.
A check with the planimeter shows a nearly identical area estimated and measured (deviation
of estimate —1 to —3%, depending on collaborator).

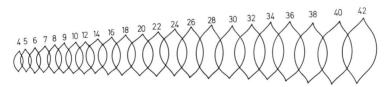

Fig. 1. Standard of beech leaf images from 4 to 42 cm² (reduced)

The material used for the estimates is leaf litter, collected during leaf-biomass
estimation in litter traps. Some damage to the leaf litter during transport and drying
inevitably occurs. We must use the leaves in a dry state because they are needed for
further chemical and caloric analysis. For L.A.I. estimation a subsample of at least
100 g (constant weight at 105° C) per period is necessary. It is divided into two
replications.

The third method is used parallel to method 2, because it was found during our studies to be a disadvantage that method 2 works well only with intact leaves. Otherwise a subjective influence through selection cannot be excluded. At least 500 discs (of 200 mm² each), preferably fresh respectively rewetted, should be punched out. Considering the smallness of the samples which are necessary, this loss of material for other investigations is acceptable. Method and evaluation work rather quickly. The published data are based on this method (Fig. 2).

Fig. 2. Perforator for leaf discs

With methods 2 and 3, the area of the whole sample for the period and/or for the year can be calculated from the leaf-area-to-leaf-weight ratio (Newbould). Because of the variations in litter structure (sun or shade leaves) with season and type, samples of all periods should be analysed. The variation in L. A. I. based on subsamples of the same period may amount to 3%. The conversion factor for dry to fresh leaf area was found after many measurements to be $1.14 \pm 0.01\%$.

## II. Preliminary Results

Some preliminary results are shown in Table 1.

Table 1. *Leaf area index (L.A.I.) of fresh leaves*

| Plot | B 1 | B 3 | B 4 |
|------|-----|-----|-----|
| 30. 4. 1967—17. 4. 1968 | 6.4 | — | — |
| 18. 4. 1968—16. 4. 1969 | 5.7 | 6.4 | 6.0 |
| 17. 4. 1969— 5. 5. 1970 | 5.6 | 6.4 | 6.2 |

In the vegetation period 1967 the beech leaf area was considerable larger than in the following ones. A similar recession in leaf area was observed on meadows (see GEYGER); some explanation is expected from the nutrient cycle investigations as biomass did not change to such a great extent. Meteorological influences can be interpreted after a further period.

With leaf litter there is the difficulty that sun and shade leaves are mixed in one trap. From observations it is possible to make only approximate conclusions regarding the main type of leaves in the trap. During our special investigations of the crown, we observed that there are certain limits for sun and shade leaves in leaf thickness. Therefore we are trying to use measurements of leaf thickness to determine the proportion of each leaf type in the traps, thus obtaining a stratified L.A.I.

## III. Methods for Spruce Needles

In spruce *(Picea abies)* forests, leaf litter cannot be used to calculate L.A.I. In this case the index is obtained by analysing subsamples of the needle biomass of the trees. Investigations are under way, along the lines of research by MØLLER (1945), SCHÖP-FER (1961) and DROSTE (1969).

## References

DROSTE ZU HÜLSHOFF, B. v.: Struktur und Biomasse eines Fichtenbestandes auf Grund einer Dimensionsanalyse an oberirdischen Baumorganen. Diss. München 1969.

MARSHALL, J. K.: Methods for leaf area measurements of large and small leaf samples. Photosynthetica 2, 41—47 (1968).

MØLLER, C. M.: Untersuchungen über Laubmenge, Stoffverlust und Stoffproduktion des Waldes. Det forstlige Forsøgsvæsen i Danmark 17 (145), 1—287 (1945).

NEWBOULD, P. J.: Methods for estimating the primary production of forests. Oxford: Blackwell, 1967.

SCHÖPFER, W.: Beiträge zur Erfassung des Assimilationsapparates der Fichte. Schriftenreihe der Landesforstverwaltung Baden-Württemberg 10, 1—127 (1961).

# C. Phenological Observations on Beech and Spruce as a Function of Climate

R. Schober and G. Seibt

## I. Methods

In 1967 the leafing and the unfolding of leaves on all stems on plots B 1 and F 2 were recorded weekly, on beeches also leaf yellowing and fall, according to features stated in Table 1:

Table 1. *Phenological phases in spruce and beech*

| Phase | Phenological observations | Phase | Phenological observations |
|---|---|---|---|
| spruce: 0 | buds resting | beech: 0 | buds resting |
| 1 | buds swollen | 1 | buds swollen |
| 2 | needle points emerging (first green) | 2 | first green |
| 3 | short shoots half unfolded | 3 | leaves half unfolded |
| 4 | short shoots fully unfolded | 4 | leaves fully unfolded and green |
| 5 | long shoots emerging | 5 | leaves begin to yellow |
| | | 6 | all leaves yellow |
| | | 7 | leaves brown |
| | | 8 | some leaves fallen |
| | | 9 | all leaves fallen |

The phenological observations were made from ground level with $10 \times$ magnifying binoculars.

After the phenological development of all 245 beech and 75 spruce trees on B 1 and F 2 had been observed in 1967, early-, middle- and late-sprouting stems in different sociological tree and diameter classes on these plots were selected for further observation:

| Plot | Tree species | Test trees with phenological beginning of vegetation period | | | Total test trees |
|---|---|---|---|---|---|
| | | early | middle | late | |
| B 1 | beech | 12 | 11 | 13 | 36 |
| F 2 | spruce | 8 | 3 | 9 | 20 |

The phenological observations of these test trees were repeated weekly in 1968 and 1969, and at shorter intervals if necessary, at the beginning and end of the vegetation period. The weekly radial increment was measured at the same time.

## II. Phenological Phases Observed in 1967

The phenological phases observed in 1967 are synchronized in Fig. 1 with the data for air temperature and rainfall.

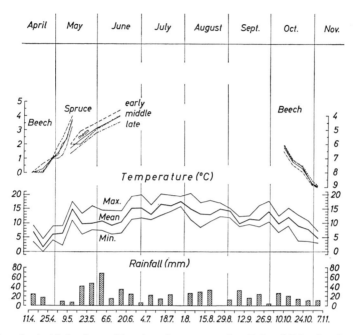

Fig. 1.  Phenological behaviour of beech and spruce on plots B 1 and F 2 related to temperature and rainfall in 1967

Fig. 1 shows that the beeches sprouted earlier and more rapidly than the spruces. Spruces showed a greater difference between early and late sprouts (about 8–13 days) than the beeches. The following observations were made with the *beech*: trees whose leaves unfolded first, yellowed first.

Frosts that might have advanced the falling of leaves were not observed.

## III. Phenological Behaviour as a Function of Temperature and Rainfall

The dependence of the phenological behaviour of the beech upon temperature and rainfall can be seen in Table 2, explaining in detail Fig. 1.

Table 2 and Fig. 1 show that in 1967 the phenological vegetation period began after a weekly average temperature of 6° C was exceeded and ended when the weekly average *air temperature* fell below 10° C.

Phenological phenomena were not clearly seen to depend on the *quantity of rain* which was thought to be due to the location of the stand, its elevation and the water capacity of the soil, which was obviously sufficient in 1967.

Table 2. *Phenological behaviour of the beech on plot B 1 as a function of temperature and rainfall in 1967*

| Week | Daytime temperature[a] | | | Rainfall[b] weekly mm | Phenological phase at *end* of examination week | | | Beginning, end and length of vegetation period |
|---|---|---|---|---|---|---|---|---|
| | weekly mean °C | minimum during week °C | maximum °C | | early-sprouting trees | medium early-sprouting and mean of all trees | late-sprouting trees | |
| 11. 4.—17. 4. | 6.9 | 3.4 | 9.1 | 24 | sporadic swelling of buds | — | — | |
| 18. 4.—24. 4. | 1.5 | —0.3 | 4.4 | 17 | swelling of buds on most trees | swelling of buds on half the trees | sporadic swelling of buds | |
| 25. 4.— 1. 5. | 6.0 | 4.0 | 8.9 | — | first green on most trees | first green on some trees | swelling of buds on all trees | |
| 2. 5.— 8. 5. | 6.4 | 2.2 | 9.0 | 9 | on all trees leaves half unfolded | on half the trees leaves half unfolded | first green on all trees | |
| 9. 5.—15. 5. | 14.9 | 11.0 | 17.5 | 7 | on all trees leaves fully unfolded | on all trees leaves fully unfolded | on most trees leaves fully unfolded | about 8. 5. 67 beginning of phenological vegetation period after a week's average temperature exceeding 6° C |
| 8. 5.— 3. 10. | | | | | vegetation period with green foliage | | | length of phenological vegetation period of examined trees 148 days. |
| 3. 10.— 9. 10. | 9.7 | 6.8 | 12.4 | 26 | yellow foliage on all trees | brown foliage on some trees | brown foliage on half the trees | about 3.10.67 end of phenological vegetation period after a week's mean temperature of 10° C |
| 10. 10.—16. 10. | 12.1 | 9.0 | 15.3 | 20 | brown foliage on all trees, sporadic leaf fall | brown foliage on all trees, sporadic leaf fall | brown foliage on all trees, sporadic leaf fall | |
| 17. 10.—23. 10. | 9.0 | 3.8 | 12.9 | 14 | partial leaf fall | partial leaf fall | partial leaf fall | |
| 24. 10.—30. 10. | 7.9 | 3.6 | 10.9 | 11 | almost total leaf fall | almost total leaf fall | almost total leaf fall | first frosts: 20.11.67 = —0,4° C |

[a] *Temperature:* average, maximum and minimum day temperature during the weeks of observation according to the data of the IBP Station on plot B 1, altitude 500 m.

[b] *Rainfall:* amount of rainfall according to data of the met. station Schiesshaus, 390 m.

Table 3. *Phenological behaviour of spruce on plot F 2 as a function of temperature and rainfall in 1967*

| Week | Daytime temperature[a] weekly mean °C | minimum during week °C | maximum during week °C | Rainfall[b] weekly mm | Phenological phase at *end* of examination week early-sprouting trees | medium early-sprouting and mean of all trees | late-sprouting trees | Beginning, end and length of vegetation period |
|---|---|---|---|---|---|---|---|---|
| 9. 5.—15. 5. | 14.9 | 11.0 | 17.5 | 7 | on half the trees half unfolded short shoots | on all trees needle points grown through | on half the trees needle points grown through | about 30. 5. 67 beginning of phenological vegetation period after an average temperature exceeding 10° C |
| 16. 5.—22. 5. | 9.6 | 6.1 | 13.4 | 41 | half unfolded short shoots on all trees | half unfolded short shoots on half the trees | on some trees half unfolded short shoots | |
| 23. 5.—29. 5. | 9.9 | 7.6 | 16.0 | 47 | totally unfolded short shoots on half the trees | half unfolded short shoots on half the trees | on most trees half unfolded short shoots | |
| 30. 5.— 5. 6. | 10.7 | 7.4 | 14.5 | 68 | totally unfolded short shoots on most trees | totally unfolded short shoots on half the trees | on all trees half unfolded short shoots | |
| 6. 6.—12. 6. | 9.2 | 6.1 | 14.4 | 15 | on some trees long shoots grown through | on most trees totally unfolded short shoots | on half the trees totally unfolded short shoots | |
| 13. 6.—19. 6. | 10.4 | 6.5 | 14.3 | 34 | on half the trees long shoots grown through | on some trees long shoots grown through | on most trees totally unfolded short shoots | first frosts: 20. 11. 67 = —0,4° C |

[a] *Temperature:* average, maximum and minimum day temperature during the weeks of observation according to the data of the IBP Station on plot B 1, altitude 500 m.
[b] *Rainfall:* amount of rainfall according to dates of the met. station Schiesshaus, 390 m.

Since the early-sprouting beeches showed green six days earlier than the late-sprouting ones, but yellowed three days later in autumn, the vegetation period of the early-sprouting trees was longer than that of the late-sprouting trees. The phenological vegetation period of the beech consisted of 148 days in the mean of all test trees.

The phenological stages of the *spruce* are also shown in Fig. 1 and are separately explained in Table 3. With spruce the phenological vegetation period may be considered to begin when the week's average temperature exceeds 10° C, thus 21 days later than that of beech. The end and hence the duration of the phenological vegetation period of spruce cannot be ascertained by optical means, as needles have a lifetime of several years; therefore dates that could be compared with those for beech cannot be given here. Continuous chemical testing of needle samples in the crowns to determine the end of the vegetation period would have required expensive constructions and these were not yet available.

# D. Structure and Timber Production of the Forest Stands

## Report on the Methods Used and Some Results

### R. Schober and G. Seibt

## I. Introduction

The present study deals with that part of primary production which is most important in the human economy: timber production.

Annual surveys were made of the standing crop at the end of the growing season, and differences were recorded, so that the increment of beech and spruce wood was ascertained for whole vegetation periods. Besides these increment measurements for entire vegetation periods, the beginning, end and extent of the weekly radial increments of beech and spruce were established. At the same time the relationship was ascertained between the extent and rhythm of this radial increment and the phenological behaviour of the measured trees and the local climate (temperature and rainfall, see chapter C).

## II. Standing Crop and Yield

### 1. Selection of Sample Plots

Sample plots sought in young, medium and mature stands of beech and spruce of the Solling Project required that the site be as homogeneous as possible and the stand evenly grown. The main condition was that age within the stands should be uniform.

The determination of age was made according to forestry records or by counting annual rings at the base of the trunks of sample trees. The years needed to achieve felling height were added.

Differences in age within the same stand were found in planted stands due to repair planting and within natural stands to seeding in different years.

The size of the plot depended on the purpose of the experiment: the main research plots B 1 und F 1 were 1 ha, the control areas were smaller, being 0.1 ha. The spruce control plot F 2 formed an exception, being 0.25 ha in order to provide a sufficient collection of stems for the research, as it was an old stand containing comparatively few trees.

Each experimental plot was marked by posts at the corners, protected by spur ditches.

Each stem on the experimental plots was numbered on the side away from the prevailing wind in weather-resistant oil paint and marked with a cross 1.3 m (breast height) above ground level. On the spruce stands of plots F 1, F 2 and F 3 where the bark was stripped by red deer, the cross was placed above the damaged area at a height of 2 m.

## 2. Stand Inventory

The *diameter of each tree* of the main experimental plots B 1 and F 1 and the control plot F 2 was measured with a girthing tape. The trees on the remaining plots were measured with a light metal calliper calibrated in millimeters. In our experience, girth measurement is a more accurate way of measuring the increment, but it takes more time and gives rather higher results than measurements made with the calliper.

For the *measurement of tree height* and the extent of the crown to determine the length of the crown, the "Blume-Leiss" trigonometrical dendrometer was used. The height of 40 stems was measured per experimental plot, except on the main plot B 1, where the height of all 245 beeches had been determined during the first survey.

## 3. Computation of the Results

The *diameter* of the stems at breast height, measured in millimeters, was rounded off to the nearest centimeter and converted into a basal area. The sum of all basal areas was divided by the number of stems to give a "mean basal area", which is considered to give mean diameter of the stand. For diameters measured at the height of 2 m, a conversion to the breast-height diameter was made according to the increase or decrease in diameter in different parts of the shaft of the stem. The height measurements were computed electronically to give a smoothed curve as a 2nd-degree parabola (cf. Schmidt, 1967). The basal area values classified in one-centimeter steps were multiplied by the smoothed height values. The sum of the products, divided by the sum of the basal areas of all stems, gives the "mean height of the stand".

Because of the known difficulties of measuring directly the *mass of standing trees*, the cubic content of the crop was estimated as usual by "mass tables" (Grundner-Schwappach, 1952). The compact wood masses listed in the tables according to height and diameter, i.e. the compact wood of the shaft and branch timber over 7 cm in diameter, were multiplied by the number of stems in each centimeter class and the products were summed. The number of stems, the sum of the basal areas and the compact wood mass of the individual stems were worked out on a 1 hectare (ha) basis, this being the unit of area used in forestry.

## 4. Comparison of Yield Values with Yield Table

Forestry yield tables are compiled on the basis of extensive study of pure stands of different tree species of the same age. These tables are generally divided according to 5 mean height classes, giving the growing stock per ha according to 5-year age steps for the growing and for the part eliminated by thinning.

The tables also contain data on mean height, mean diameter, stem density and basal area, as well as the stand form factor and the increment. The yield tables enable estimates to be made of the stand volume and compared with the survey results obtained from the stand inventory. This brings out the association of yield values with site conditions and type of management.

## 5. Results of the First Forest Inventory in Spring 1967

As an example of the assessment of the forest yield in the sample plots of the Solling project, the results of the forest inventory made in spring 1967 and their comparison with the yield table (Wiedemann and Schober, 1957) are given in Table 1.

Table 1. *Yield values of the sample plots in spring 1967 and comparison with the yield table by* WIEDEMANN-SCHOBER (1957)

| Compartment and yield table used for comparison | Plot and tree species | Area ha | Age years | Number of stems | Mean height m | Yield class | Basal area m² | Mean diameter (Ø) cm | Total volume >7cm Ø m³ | Performance grade[a] m³ |
|---|---|---|---|---|---|---|---|---|---|---|
| Comp. 51 | B 1 beech | 1.0 | 120 | 245 | 25.4 | III.2 | 26.43 | 37.1 | 348.2 | 5 |
| yield table | | | 120 | 253 | 26.3 | III.0 | 23.4 | 34.3 | 310.0 | |
| heavy thinning | | | diff. | −3% | | | +13% | +8% | +8% | |
| Comp. 91 | B 3 beech | 0.1 | 78 | 1190 | 19.8 | III.0 | 23.2 | 15.8 | 218.5 | 5 |
| yield table | | | 80 | 780 | 19.9 | III.0 | 23.6 | 19.6 | 228.0 | |
| heavy thinning | | | diff. | +53% | | | −2% | −19% | −4% | |
| Comp. 27 | B 4 beech | 0.1 | 57 | 3620 | 15.0 | II.9 | 26.87 | 9.7 | 158.7 | 6 |
| yield table | | | 55 | 2413 | 14.1 | III.0 | 24.2 | 11.3 | 147.0 | |
| moderate thinning | | | diff. | +50% | | | +11% | −14% | +8% | |
| Dassel Comp. 28[b] | F 1 spruce | 1.0 | 85 | 595 | 24.4 | II.4 | 42.79 | 30.3 | 510.1 | 9 |
| yield table | | | 85 | 706 | 26.6 | II.0 | 43.1 | 27.9 | 561.0 | |
| moderate thinning | | | diff. | −16% | | | −1% | +11% | −9% | |
| Comp. 83 | F 2 spruce | 0.25 | 113 | 300 | 31.0 | II.4 | 36.19 | 39.2 | 524.8 | 9 |
| yield table | | | 115 | 268 | 32.5 | II.0 | 36.0 | 41.4 | 527.0 | |
| heavy thinning | | | diff. | +12% | | | +1% | −5% | ±0% | |
| Comp. 25 | F 3 spruce | 0.1 | 39 | 1490 | 15.5 | I.3 | 31.83 | 16.5 | 255.4 | 10 |
| yield table | | | 40 | 1416 | 17.4 | I.0 | 30.4 | 16.5 | 272.0 | |
| heavy thinning | | | diff. | +5% | | | +5% | ±0% | −6% | |

[a] Mean annual increment (total yield volume >7 cm Ø) at the age of 100.
[b] All other compartments belong to district Neuhaus.

The yield values shown in Table 1 for three experimental plots reveal that spruce stands have more stems of greater height and diameter compared to beech stands of approximately the same age, and consequently provide more compact wood and a higher yield class. Compared to the yield tables for north-west Germany, the stands of most sample plots show exceedingly "heavy thinning", though the young beech stands of plot B 4 are more densely stocked than moderate thinning in the yield table; also the spruce stands were planted in normal spacing (about 4500 plants per ha). Since the stands have not been heavily thinned so far, there must be other reasons for the poor growing stock of the old stands on the sample plots. First of all, there is bark damage by red deer, causing quick decay of the beech timber, slower decay of spruce timber and a loss of trees. Since the strongest trees in young stands are stripped first, it is the stand members with the best increment that are lost. Thinning then eliminates not the badly shaped or less well-grown trees, but those with the most bark peeling damage, thus reducing the quality of the timber.

Plot F 1 is a special case, since the stand planted on old pasture land probably grew very dense owing to its origin and, contrary to expectation, stripping did not destroy the trees.

## III. Structure and Quality of the Stands

### 1. Method

During the first stand survey in spring 1967, every tree was allocated to one of 5 classes in its sociological position (Tree Classification of the Forest Experiment Stations of 1902). Furthermore, quality factors or defects and damage to shafts and crowns were noted for every stem (see SCHOBER, 1954, and WIEDEMANN, 1962).

### 2. Result of the Survey of the Beech Stand B 1

The result of judging each stem for tree class, stem shape, crown shape and branchiness on the stand of plot B 1 is shown in Table 2.

The investigations show that according to the *crown structure* the beech stand B 1 is mainly *one-storied:* 84% of the trees form the upper crown layer of the dominant trees while only 8% of the lower crown layers belong to the intermediate or understory class.

On stem quality, the experimental stand can only be considered as *average to poor*, since only 18% of its stems show straightness of bole, $^3/_4$ being crooked and bent and $^1/_5$ showing forking and branchiness.

The crown structure shows variety of forms which is typical for the beech; only 58% of the stems have normal or large crowns.

### 3. Stem and Crown Distribution

#### a) Special Method

The distribution of stems and crowns over the area occupied by the stand is important in order to judge whether the growing space is fully utilized. The crown size also enables estimates to be made of the increment. For this purpose, on plots

Table 2. *Tree class, stem shape and crown shape according to diameters in a 120-year-old beech stand: plot B 1, survey 1967*

| Breast height diameter cm | Tree class | | | Stem | | | | Crown | | | |
|---|---|---|---|---|---|---|---|---|---|---|---|
| | dominant (no. of trees) | inter-mediate | under-storied | straight | crooked, bent | forked | branchy (addi-tional) | normal or large | one-sided | poor | cramped |
| 16—20 | — | — | 8 | — | — | — | — | — | — | 8 | — |
| 21—25 | 1 | 8 | 10 | 1 | 9 | — | 2 | — | 3 | 16 | — |
| 26—30 | 24 | 11 | — | 6 | 29 | 1 | 10 | 8 | 14 | 13 | — |
| 31—35 | 36 | 2 | 1 | 4 | 36 | 1 | 10 | 18 | 13 | 5 | 3 |
| 36—40 | 64 | — | — | 12 | 51 | 4 | 5 | 49 | 12 | 1 | 2 |
| 41—45 | 49 | — | — | 6 | 42 | 8 | 3 | 37 | 10 | 1 | 1 |
| 46—50 | 21 | — | — | 11 | 10 | 2 | — | 21 | — | — | — |
| 51—55 | 8 | — | — | 3 | 4 | 3 | — | 8 | — | — | — |
| 56—60 | — | — | — | — | — | — | — | — | — | — | — |
| 61—65 | 2 | — | — | — | 1 | 1 | — | 2 | — | — | — |
| Total: number | 205 | 21 | 19 | 43 | 182 | 20 | 30 | 143 | 52 | 44 | 6 |
| % | 84 | 8 | 8 | 18 | 74 | 8 | — | 58 | 21 | 18 | 3 |

B 1 and F 1 location of each stem on the ground was measured, also the projection of the outer edge of the crown. The 1 ha plots were divided into map squares of 5 × 5 m, marked at the corners by stakes. Using the corner plots as coordinates, the position and projection of edges of the crown of each stem were measured by prismatic compass and steel measuring tape. The crown edges were measured with a crown reflector (Seibt, 1968) specially constructed for wide-crowned beeches, since the usual crown plummets are suited only for the small crowns of conifers.

From the measurements a stem and crown map was drawn on a scale of 1:200. The crown area was determined with a polar planimeter.

### b) Stem and Crown Distribution of Plot B 1

In order to carry out the classification of the crown levels and size, the field measurements had to be made in the early spring of 1967, before the beeches came into leaf. The stem and crown map has already been published (Ellenberg, 1967).

Table 3 shows how the area sheltered by the beech crowns is distributed.

Table 3. *Crown area on plot B 1 in the year 1967*

|  | m² | % of total stand area |
|---|---|---|
| A. Crown projection area | | |
| 1. of stems within the experimental plot: | 8252.10 | 82 |
| 2. of stems on the edge of the experimental plot or outside it, with part of their crowns extending into it: | 1083.52 | 11 |
| Total crown projection area: | 9335.62 | 93 |
| B. Area not covered by crowns: | 664.38 | 7 |
| C. Total area with and without crown cover: | 10000.— | 100 |
| Of the total crown projection plot (A.1 + A.2): | | |
| a) was covered by 1 crown: | 8347.60 | 89 |
| b) was covered by 2 crowns: | 975.55 | 11 |
| c) was covered by 3 crowns: | 12.47 | — |
| Total: | 9335.62 | 100 |

This summary shows the comparatively even density of the crown cover, filling 93% of the crown space, while in addition over 11% of the crown area two and, in a few places, three trees had developed their crowns, one over the other. As a result of this relatively high crown density, the growing stock in terms of basal area and compact wood, shown in Table 1, exceeds the value for heavy thinning given in the yield table.

## IV. Timber Increment

### 1. Method of Increment Research

The increment in timber volume was determined by repeated surveys of the growing stock on the experimental plots. Table 4 shows dates and findings of surveys made to determine the increment in the stands of the various plots.

Table 4. *Yield survey carried out in the stands of experimental plots*

| Plot: | B₁ | B₃ | B₄ | F₁ | F₂ | F₃ |
|---|---|---|---|---|---|---|
| Area (ha): | 1.0 | 0.1 | 0.1 | 1.0 | 0.25 | 0.1 |
| *February/April 1967* | | | number of stems | | | |
| double tallying | 245 | 119 | 362 | | | 149 |
| girth measurement | 245 | | | 595 | 75 | |
| height measurement | 245 | 40 | 40 | 40 | 40 | 40 |
| stem distribution and | | | | | | |
| crown mapping | 245 | | | 595 | | |
| *March 1968* | | | | | | |
| girth measurement | 244 | | | 595 | 75 | |
| *March 1969* | | | | | | |
| double tallying | | 119 | 362 | | | 149 |
| girth measurement | 243 | | | 595 | 75 | |
| height measurement | 40 | 40 | 40 | 40 | 40 | 40 |

Table 4 shows that in the large experimental plots B 1, F 1 and F 2 annual girth measurements were made of all stems to ascertain the increment in diameter of basal area, and height measurements were made after two years. The yield survey of the 0.1 ha check plots B 2, B 3 and F 3 made at one-year intervals are not appropriate for exact determination of increment, as the area is too small.

## 2. Stand Increment of the Larger Experimental Plots

Table 5 shows that in 1968 the *basal area increment* was higher than in 1967, apart from the increment on the spruce stand of sample plot F 2, which was about the same

Table 5. *Increment of basal area and timber on plots B 1, F 1 and F 2 and according to the Wiedemann-Schober yield table*

| Plot: | | | B 1 | F 1 | F 2 |
|---|---|---|---|---|---|
| Tree species: | | | beech | spruce | spruce |
| Age in 1967 (years): | | | 120 | 85 | 113 |
| Size of plot, ha: | | | 1.0 | 1.0 | 0.25 |
| Actual annual increment: | year | | | | |
| basal area | 1967 | m² | 0.57 | 0.94 | 0.64 |
| | 1968 | m² | 0.75 | 1.06 | 0.63 |
| Average | 1967/68 | m² | 0.66 | 1.00 | 0.64 |
| Actual annual increment: | | | | | |
| timber (Average) | 1967/68 | m³ | 16.8 | 17.2 | 10.7 |
| *Yield table values for comparison* | | | | | |
| Type of thinning: | | | heavy | moderate | heavy |
| Yield class: | | | III | II | II |
| Age (years): | | | 120/125 | 85/90 | 115/120 |
| Averaged annual increment basal area | | m² | 0.48 | 0.64 | 0.42 |
| Averaged annual increment timber | | m³ | 7.7 | 10.6 | 7.4 |

in both years. At approximately the same age as the beech stand B 1, the spruce stand F 2 had a higher increment of basal area than the beech in 1967 and a lower one in 1968, while the average of both years was about the same. Reference to the yield tables shows that this similarity is quite normal, as is the higher increment of basal area and timber in the younger spruce stand compared to the older one. All basal area increments on the experimental plots were higher than those given in the yield table, though the stands (see Table 1) are normally stocked according to the yield tables, only the beech stand B 1 being overstocked. The *timber volume increments* calculated from the stand inventory, which was repeated after two years, though shown in Table 5, cannot be considered reliable. The usual methods of calculating forestry yields on the basis of volume tables permit a reliable statement of volume increment only for periods of ten or more years. In the present case, the annual increment of plots B 1 and F 1 in particular is higher than the average. We hope further research and collaboration with projects A, B, E, S, U and X will help to find the reasons for this interesting fact.

# References

Ellenberg, H.: International Biological Programme. Contributions of the German Federal Republic. Bad Godesberg: German Research Association 1967.

Grundner-Schwappach: Massentafeln zur Bestimmung des Holzgehalts stehender Waldbäume und Waldbestände. 10. Aufl., Berlin 1952.

Kunze, M.: Hilfstafeln für Holzmassenaufnahmen. 5. Aufl., Berlin: Parey 1938.

Mitscherlich, G.: Sortenertragstafel für Fichte. Mitt. Forstwirtsch. 569—583 (1939).

Schmidt, A.: Der rechnerische Ausgleich von Bestandshöhenkurven. Forstw. Cbl. 370—382. (1967).

Schober, R.: Baumklassenansprache auf Versuchsflächen. Unpublished 1954.

Seibt, G.: Ein neuer Spiegel für die Gewinnung von Kronenkarten. Allg. Forstz. 29 (1968).

Wiedemann, E.: Anweisung für die Aufnahme und Bearbeitung der Versuchsflächen der Preußischen Forstlichen Versuchsanstalt. Neudamm: Neumann 1926.

— Schober, R.: Ertragstafeln wichtiger Holzarten bei verschiedener Durchforstung. Hannover: M. H. Schaper 1957.

# E. Estimation of Biomass of Forests

H. Heller

## I. Introduction

The methods used in the biomass studies in the Solling forests correspond to those reviewed by Newbould (1967: IBP Handbook No. 2). Their object is to obtain correlations between "destructive" and "non destructive" measurements. Our program starts from the results obtained by the team of the Niedersächsische Forstliche Versuchsanstalt (see Schober and Seibt, chapter D); it is combined with the measurements of Eber (see chapter G) on ground vegetation and the work of Meyer and Göttsche (see chapter F) on the distribution of fine roots.

The following components are considered separately:

leaves, fruits, flowers and bud scales;

stem and branches;

roots.

All weights are measured at 105° C (dried until constant weight is obtained). Losses of volatile materials are negligible (see Runge, chapter L).

## II. Leaves etc.

Litter traps (surface area 0.5 m²) serve as collectors for bud scales, flowers, fruits and leaves. The funnels are distributed randomly on the paths of the sample plot. For the number of traps in the six sample areas, see Table 1; it also shows the percentage of the area used for catching and the variation of results between the traps for one year (May—April).

The bags are normally emptied after 2 or 3 weeks, in autumn once a week. The material is sorted, dried and weighed and a calculation is made of transfer from

Table 1. *Biomass in traps (not ashfree)*

|  | no. of traps | % of sample area | Biomass 1967 (t/ha/yr) total  leaves | | Variation coefficient between traps (%) | Biomass 1968 (t/ha/yr) total  leaves | | Variation coefficient between traps (%) | Biomass 1969 (t/ha/yr) total  leaves | |
|---|---|---|---|---|---|---|---|---|---|---|
| B 1 | 20 | 0.1 | 4.0 | 3.4 | 9 | 3.8 | 3.2 | 5 | 3.4 | 3.0 |
| B 3 | 10 | 0.5 | incomplete | | | 3.7 | 3.4 | 17 | 3.6 | 3.2 |
| B 4 | 10 | 0.5 | incomplete | | | 3.7 | 3.3 | 10 | 3.4 | 3.0 |
| F 1 | 20 | 0.1 | 2.5 | | 7 | 2.8 | | 15 | | |
| F 2 | 25 | 0.5 | incomplete | | | 2.2 | | 11 | | |
| F 3 | 10 | 0.5 | incomplete | | | 2.2 | | 52 | | |

canopy to forest floor in kg/ha · day, and summed, converting to metric tons/ha/ year.* Undervaluation due to leaf weight changes during the year and secondary producers is not yet compensated for.

Furthermore small twigs and branches are gathered once a year on five areas, (10×10 m) in the sample areas B 1 and F 1. This investigation is only an additional one, because values thus obtained are already calculated through tree analysis (sect. III).

## III. Stem and Branches

According to the range of diameter at breast height (1.30 m) of the trees in the plot, sample trees are felled near the measurement area, beech during winter, spruce at the end of the growing season.

The felled stems are measured with the bark and divided into timber and brush-wood (<7 cm ⌀).** The timber is cut into 1.10 m lengths and weighed in the fresh state. Two sections 5 cm thick are sawn off, taken to the laboratory, dried and used for further investigations (chemical, caloric, increment studies). From the ratio of fresh weight: dry weight of the section the dry weight of the total stem can be calculated. 1% is added for losses as sawdust. Percentage of bark is derived from subsamples for the different diameter groups. There is a good correlation between bark and stem dry weight.

As our investigations are part of a pilot project, all brushwood from each sample tree was collected, measured, sorted, dried and weighed. This complete utilization should offer the possibility after final calculation of making more precise recommendations for subsampling in further production studies. Basic material on this subject is scarce. With beech, the sorting was done by cm-classes, separately for each main branch of the tree, except for the annual production of buds and twigs. Brushwood from spruce was collected separately for each whorl. This material was sorted into three groups: current annual production, rest of green parts (sometimes further divided by years), dead plant matter. Within the green material, needles were removed from the twigs.

The data for single sample trees will allow the biomass of the stands to be calculated by means of regression between this destructive sampling and recurrent non-destructive measurements of the stand (diameter at breast height, total height of the tree). As the differences between these annual increment measurements are rather small, it is difficult to obtain reliable results by the conventional measurements. Therefore it is necessary to correct the values thus obtained by tree ring research at the end of the study period (1972).

A calculation based on 27 trees (*Fagus silvatica* L.) gives the following preliminary tree biomass (including ash) for the research plots (Table 2).

The values are derived from the logarithmically transformed equation $y = -0.3518326 + 0.9356902 \, x$ $(r = 0.994)$, where $y = $ log total dry weight in kg, $x = \log \frac{1}{4} \, d^2 \cdot \pi \cdot h$ ($d = $ diameter at breast height, $h = $ total height of tree). Problems about applicability and range of validity of this treatment will be discussed in a separate paper.

* Grams/m² ≙ t/km² ≙ 10 kg/ha ≙ 0.01 t/ha; 1 t/ha ≙ 100 g/m² ≙ 10mg/cm².
** ⌀ means diameter at breast height.

Table 2. *Biomass of beech trees* (t/ha)

| Plot | B1 | B3 | B4 |
|------|-----|--------|--------|
| April 1967 | 184 | 140 | 141 |
| April 1968 | 188 | n. det. | n. det. |
| April 1969 | 200 | 154 | 153 |

## IV. Roots

Root biomass is estimated by a combined program carried out together with
MEYER and GÖTTSCHE (chapter F). Our studies include only roots $> 0.5$ cm $\varnothing$. Their
mass is calculated on the basis of excavations. Roots of some sample trees are dug
out, washed and dried (OGAWA et al.). By adding MEYER's values, we think a fairly
good estimate will be obtained in future summaries.

## References

NEWBOULD, P. J.: Methods for estimating the primary production of forests. Oxford:
Blackwell 1967.

OGAWA, H., YODA, K., OGINA, K., KIRA, T.: Comparative ecological studies on three main
types of forest vegetation in Thailand II. Plant biomass. Nature and life in southeast
Asia, **4**, 49—80 (1965).

# F. Distribution of Root Tips and Tender Roots of Beech

F. H. Meyer and D. Göttsche

## I. Methods

Various methods are utilized to investigate the root systems of forest trees (Weller, 1964; Röhrig, 1966; Meyer, 1967; Köstler et al., 1968):

1. The length of roots in different root classes. These classes are usually differentiated according to root diameter:

| | | |
|---|---:|---|
| large roots | >20 | mm |
| medium roots | 5—20 | mm |
| small roots | 2— 5 | mm |
| fine roots | 0.5— 2 | mm |
| finest roots | < 0.5 | mm |

2. The surface of different root classes;
3. The weight of different root classes;
4. The number of root tips per volume of soil (the root tips can be subdivided according to the type of mycorrhiza, growing activity etc.);
5. The number of root tips per cm length of fine root.

This investigation covered only weight of fine and finest roots, weight of small roots, and number of root tips per volume of soil and took place in a pure *Fagus silvatica* stand near research plot B 1.

To estimate the biomass of small roots, a sample block $30 \times 80$ cm was carefully excavated to a depth of 80 cm. From this block all small roots were removed, washed, oven-dried at 105° C and weighed. The values given below refer to one $30 \times 80$ cm block taken at a distance of 2 m from a codominant tree in May 1969.

To measure the biomass of fine and finest roots and to count the root tips per volume of soil, 100 ml samples were collected from different subhorizons of the soil profile with the aid of a steel cylinder (6.5 cm diameter and 4 cm long) in May 1969.

Although the fine and finest roots are nearly evenly distributed between trees (Grosskopf, 1950; Kern et al., 1961; Karizumi, 1968), the cylinder samples were taken 2 m from a codominant tree so that the vertical distribution of fine and finest roots can be compared with those of small roots at this distance.

For the fermentation layer (F) and the humus layer (H) their complete depth was sampled, but the B horizon was investigated only at intervals of ca. 10 cm. The values of the biomass of fine and finest roots in the B horizon were derived from these samples (length 10 cm).

Since it was impossible to investigate all the samples immediately, they were soaked in a solution of 600 ml ethanol (96%), 180 ml formol, 30 ml acetic acid and 1190 ml water.

The soil was washed from the fine and finest roots in the 100 ml samples, and the number of intact root tips was determined under a stereomicroscope. Samples from the F and H layers containing more than 1000 root tips per 100 ml were divided, only half or a quarter being analysed. Each value in Table 1 is derived from 8 to 12 parallels.

After the number of root tips had been counted the samples were used to calculate the biomass of the fine and finest roots. The usual procedure, simple drying at 105° C, proved to be inadequate since mineral particles, often attached by mycorrhizal fungi, adhered to the roots which therefore could not be washed completely clean.

Hence samples of fine and finest roots were oven-dried at 105° C, weighed and ashed in a muffle oven at 600° C for 4 hours. The difference between the dry weight of the roots and the weight of ashes indicates the amount of organic matter. The biomass of fine and finest roots was obtained by adding to this difference an average amount of ashes of clean roots expressed as a percentage of the oven-dry weight of clean roots. To get clean roots, samples were taken from the F layer (containing predominantly decaying leaves) and all the adhering particles carefully removed from the roots under a stereomicroscope.

The biomass obtained by the above procedure was 5 to 60% less than that obtained only with drying at 105° C.

## II. Results and Discussion

### 1. Number of Root Tips per 100 ml of Soil

The number of root tips per volume of soil, or the weight of unsuberized finest roots, is a very sensitive indicator of soil properties, as has been shown for orchard trees (WELLER, 1964) and for forest trees (MEYER, 1964 and 1967; McQUEEN, 1968). Both give a hint of the absorption capacity of root systems, for most of the mineral nutrients are absorbed in the region of finest roots or mycorrhizas (MELIN, 1959; HARLEY, 1959).

Table 1. *Vertical distribution of root tips in the beech plot B 1*
(in parenthesis the 95% confidence limits)

| Layer | Depth in cm | Number of intact root tips per 100 ml of soil | |
|---|---|---|---|
| AF$_2$ | 2— 3.5 | 6,259 | (± 1,063) |
| AH | 3.5— 5 | 2,523 | (± 665) |
| Aeh | 5— 7 | 961 | (± 431) |
| B | 13—17 | 439 | (± 95) |
| B | 19—23 | 309 | (± 50) |
| B | 29—33 | 328 | (± 61) |
| B | 35—39 | 346 | (± 66) |
| B | 43—47 | 194 | (± 42) |
| B | 52—56 | 123 | (± 95) |
| B | 61—65 | 67 | (± 82) |
| B | 72—76 | 75 | (± 57) |
| B | 83—87 | 13 | (± 15) |

Depending on the soil properties, the number and vertical distribution of root tips differs greatly (Meyer, 1967): The better the site and biological conditions, the more evenly are the root tips distributed within the profile. The concentration of root tips in the A horizon and the simultaneous decrease in the B horizon may be a sign of the deterioration of nutrient status and biological conditions of the site (cf. Fig. 1). The maximum number of root tips in the A horizon of beech stands varies between 556 (eutrophic brown earth) and 45,600 (podzol). The number of root tips for the beech plot in the Solling is given in Table 1.

As shown in Table 1, the $AF_2$ (lower fermentation layer) has the highest concentration of root tips. On the other hand the B horizon is poor in root tips.

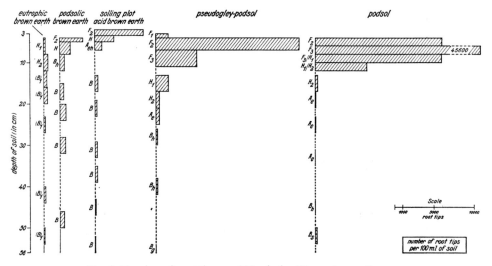

Fig. 1. Number of root tips per 100 ml of soil in various soil types.
Nomenclature of horizons and subhorizons (= layers) according to W. L. Kubiena (Bestimmungsbuch und Systematik der Böden Europas, Stuttgart: F. Enke 1953). Kubiena divides the A horizon (top soil) into the following layers: AL (= L) = litter; AF (= F) = fermentation layer; AH (= H) = humus layer; Ae = eluvial horizon

Since the root tips are distributed evenly within the F and H layers and between 3 and 47 cm depth in the B horizon, the variation between parallel cylinder samples is relatively low. But below 52 cm the fine and finest roots are distributed very unevenly in the soil. Some cylinder samples contained none and others several hundred root tips, so that wide 95% confidence limits result.

It is interesting to compare the density and distribution of root tips in a beech stand of the Solling with those in other beech stands (see Fig. 1). The number of root tips in the A horizon of the Solling plot are typical of slightly unfavourable sites. According to its rooting properties, the Solling plot might be inserted between pseudogley-podzol and podzolic brown earth. It differs from the podzolic brown earth by a higher number of root tips in the A horizon and a lower number in the B horizon. The reverse is true in a comparison with the pseudogley-podzol.

As mentioned above, the increase of root tips in the A horizon and their simultaneous decrease in the B horizon is characteristic of deterioration in site conditions. Thus the plot B 1 fits quite well in the sequence from podzolic brown earth to podzol.

## 2. Biomass of Finest, Fine and Small Roots

The values for the small roots are unreliable, for only one soil block was taken at a distance of 2 m from a codominant tree. Nevertheless, values for the small roots can be compared with those for the fine and finest roots at this distance. Table 2 indicates that the distribution of the fine and finest roots and that of the small roots is quite different.

Table 2. *The vertical distribution of biomass of finest, fine and small roots*
(values in mg per 100 ml of soil)

| Finest and fine roots | | | Small roots | | |
|---|---|---|---|---|---|
| layer | depth in cm | biomass | layer | depth in cm | biomass |
| AF$_2$ | 2 — 3.5 | 342.4 | AF$_2$ | 2 — 3.5 | 99.9 |
| AH | 3.5— 5 | 226.7 | AH | 3.5— 5 | 39.3 |
| Aeh | 5 — 7 | 70.0 | Aeh, B | 5 —15 | 150.4 |
| B | 13 —17 | 46.9 | | | |
| B | 19 —23 | 44.8 | B | 15 —30 | 50.1 |
| B | 29 —33 | 28.7 | | | |
| B | 35 —39 | 27.7 | | | |
| B | 43 —47 | 13.5 | B | 30 —50 | 60.3 |
| B | 52 —56 | 8.8 | | | |
| B | 61 —65 | 2.9 | | | |
| B | 72 —76 | 3.1 | B | 50 —80 | 7.3 |
| B | 83 —87 | 0.7 | | | |

Small roots have their best development between 5 and 15 cm. A second smaller peak lies in the AF$_2$. There were proportionally more small roots than fine and finest roots in the B horizon. The fine and finest roots dominate in the F and H layers. In the F layer they gave a biomass of 342.4 mg per 100 ml of soil, but in the deeper layers the biomass diminished gradually.

In Table 2 values are given only for living roots. The finest and fine roots contained a high proportion of dead roots, especially in the lower B horizon. The percentage of dead fine and finest roots increases from approximately 20% in the F layer to approximately 90% in the lower B horizon. This increase might, among other reasons, be associated with the lower rate of decomposition.

From the values for the biomass of small, fine and finest roots per 100 ml of soil, a calculation per ha gives the following results for the biomass of the whole profile:

finest and fine roots:  2,552 kg/ha

small            :  3,888 kg/ha.

# References

Grosskopf, W.: Bestimmung der charakteristischen Feinwurzel-Intensitäten in ungünstigen Waldbodenprofilen und ihre ökologische Auswertung. — Mitt. Bundesforschungsanst. Forst- u. Holzwirtsch., Reinbek **11**, 1—19 (1950).

Harley, J. L.: Biology of Mycorrhizae. London: Leonard Hill 1959.

Karizumi, N.: Estimation of root biomass in forests by the soil block sampling. — In: Methods of productivity studies in root systems and rhizosphere organisms (Ghilarov, M. S., V. A. Kovda et al., Eds.), pp. 79—86 Leningrad: Nauka 1968.

Kern, K. G., Moll, W., Braun, H. J.: Wurzeluntersuchungen in Rein- und Mischbeständen des Hochschwarzwaldes. Allg. Forst- u. Jagdztg. **132**, 241—260 (1961).

Köstler, J. N., Brückner, E., Biebelriether, H.: Die Wurzeln der Waldbäume. Hamburg: P. Parey 1968.

McQueen, D. R.: The quantitative distribution of absorbing roots of *Pinus silvestris* and *Fagus sylvatica* in a forest succession. Oecol. Plant. **3**, 83—99 (1968).

Melin, E.: Mycorrhiza. In: Handbuch Pflanzenphysiol. (W. Ruhland, Ed.), Bd. XI, pp. 605—638. Berlin-Göttingen-Heidelberg: Springer 1959.

Meyer, F. H.: The role of the fungus *Cenococcum graniforme* (Sow.) Ferd. et Winge in the formation of mor. In: Soil Micromorphology (A. Jongerius, Ed.), pp. 23—31. Amsterdam: Elsevier 1964.

— Feinwurzelverteilung bei Waldbäumen in Abhängigkeit vom Substrat. Forstarch. **38**, 286—290 (1967.)

Röhrig, E.: Die Wurzelentwicklung der Waldbäume in Abhängigkeit von den ökologischen Verhältnissen. Forstarch. **37**, 217—229; 237—249 (1966).

Weller, F.: Vergleichende Untersuchungen über die Wurzelverteilung von Obstbäumen in verschiedenen Böden des Neckarlandes. — Arb. Landwirtsch. Hochsch. Hohenheim **31**, 1—181 (1964).

# G. The Primary Production of the Ground Vegetation of the Luzulo-Fagetum

W. Eber

## I. Introduction

The primary production and seasonal variation of the biomass of the herb layer was studied in a *Luzulo-Fagetum* around the research plot B 1 of the Solling Project. Here the ground vegetation is very sparse, and its distribution depends mainly on the quantity of light reaching the ground. Darker parts of the stand are almost bare of plants, whereas under larger canopy holes a relatively dense vegetation is found. The ground vegetation consists mainly of *Luzula albida, Oxalis acetosella, Deschampsia flexuosa,* seedlings of *Fagus silvatica* and *Polytrichum attenuatum.* The other species contribute only a negligible proportion to biomass and production of the stand and therefore have not been studied.

At the present state of the investigations, the data and their evaluation are still incomplete, so that only a provisional survey can be presented.

## II. Method

As the number of species in this stand is very small, the individual plant method (Newbould, 1967) is most suitable for estimating biomass and production. The dry weight of individual plants, shoots or leaves was determined at monthly intervals and combined with density data for conversion to an area basis. The sum of the differences of maximum and minimum biomass of each species gives an estimate of net production.

The samples were taken near to the research plot B 1. Each sample consisted of ten plants *(Fagus),* or a great number of shoots *(Luzula, Deschampsia)* or leaves *(Oxalis).* In 1968 only three parallel samples had been taken; in 1969 this number was increased to ten. Each sample and species was analysed separately. The harvested material was sorted into subterranean components, shoots, leaves and flowers or fruits respectively. In the case of *Fagus* it proved useful to divide into seedlings which had come up in the current year and older plants. The material was washed and dried at 105° C.

For conversion to an area basis, the number of individual plants, shoots or leaves at each sample date was counted in one of the hundred $10 \times 10$ m sub-plots of the main beech sample plot B 1. A determination of the density data for the whole hectare area once a year enables the biomass of the sample plot B 1 to be calculated.

## III. Some Results

The seasonal variation of the biomass is shown in Fig. 1 and in Table 1. The data are still incomplete, as *Polytrichum* has not yet been studied and the data for *Deschampsia* are not satisfactory. The variation of the above-ground biomass is so strong that even in 1968, when only three parallel samples had been taken, the trend of these

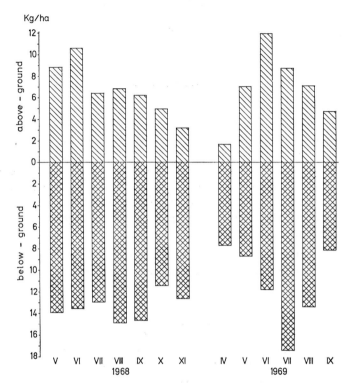

Fig. 1. Seasonal changes of the biomass (kg/ha) of the herb layer (without *Deschampsia* and *Polytrichum*)

changes could be recognized. Regarding the below-ground components, however, this is only possible with the 1969 data, although the extraordinarily high variance of parallel samples considerably limits the reliability of the results.

The minimum values of the above-ground biomass for all species were found in the early spring before growth begins. The maximum in June is caused by the immense number of *Fagus* seedlings which have come up since May. But this high proportion is rapidly decreased by the severe attacks mainly of aphids and several species of beetles. The biomass of *Luzula* and *Oxalis* in 1968 continued to increase until the late summer. In 1969 this increase was stopped in July by an extreme drought. The changes in the subterranean biomass are similar to those of the above-ground parts.

The value of annual production can be calculated approximately as the sum of the difference between maximum and minimum biomass for each species. In the case of

Table 1. *Seasonal variation of the biomass (kg/ha) of the ground vegetation near plot B 1*

| | above-ground parts | | | | below-ground parts | | | |
|---|---|---|---|---|---|---|---|---|
| | *Luzula* | *Oxalis* | *Fagus* | total | *Luzula* | *Oxalis* | *Fagus* | total |
| *1968* | | | | | | | | |
| 3. 5. | 1.37 | 0.34 | 7.15 | 8.86 | 10.78 | 1.71 | 1.42 | 13.91 |
| 5. 6. | 2.58 | 0.74 | 7.25 | 10.57 | 9.38 | 1.62 | 2.57 | 13.57 |
| 3. 7. | 2.64 | 1.11 | 2.61 | 6.37 | 9.64 | 2.17 | 1.12 | 12.93 |
| 6. 8. | 3.50 | 1.54 | 1.74 | 6.78 | 10.76 | 3.13 | 1.00 | 14.89 |
| 2. 9. | 3.45 | 1.24 | 1.50 | 6.19 | 11.37 | 2.28 | 0.96 | 14.60 |
| 3. 10. | 2.73 | 1.15 | 1.09 | 4.96 | 7.96 | 2.48 | 0.91 | 11.36 |
| 5. 11. | 1.71 | 0.49 | 0.97 | 3.18 | 9.91 | 1.84 | 0.82 | 12.57 |
| *1969* | | | | | | | | |
| 10. 4. | 1.02 | 0.17 | 0.44 | 1.64 | 6.29 | 0.86 | 0.54 | 7.70 |
| 12. 5. | 2.20 | 0.52 | 4.26 | 6.98 | 6.16 | 1.42 | 1.10 | 8.68 |
| 10. 6. | 5.23 | 0.65 | 6.06 | 11.93 | 8.66 | 1.48 | 1.59 | 11.74 |
| 8. 7. | 5.49 | 0.89 | 2.34 | 8.72 | 14.61 | 1.57 | 1.19 | 17.38 |
| 15. 8. | 5.33 | 0.50 | 1.24 | 7.07 | 11.53 | 1.05 | 0.83 | 13.41 |
| 18. 9. | 3.88 | 0.44 | 1.39 | 5.71 | 6.23 | 1.01 | 0.92 | 8.16 |
| 18. 10. | | 0.44 | | | | 0.79 | | |

*Fagus*, whose biomass decreases while weight of some individual plants continues to increase until the end of the growing period, the production had to be calculated from the monthly increment of the individual plants.

The values thus obtained for 1969 are shown in Table 2. The value for the below-ground production of *Luzula* probably is too high.

Table 2. *Production of the herb layer (kg/ha) of the Luzulo-Fagetum in 1969* (without *Polytrichum*)

| | above-ground | below-ground | total |
|---|---|---|---|
| *Fagus silvatica* (seedlings) | 6.1 | 1.5 | 7.6 |
| *Luzula albida* | 4.5 | 8.4 | 12.9 |
| *Oxalis acetosella* | 0.7 | 0.7 | 1.4 |
| *Deschampsia flexuosa* | 2.0 | 1.0 | 3.0 |
| total | 13.3 | 11.6 | 24.9 |

Although the reliability of the data for 1968 is considerably limited by the small number of parallel samples, it is obvious that the values of biomass and production will differ from year to year. Therefore it will be the task for the following years to determine the range of variation and the influence of ecological factors on growth and production of the herb layer.

To make the values as exact, as possible, the accuracy of sampling will be improved. The dead material of *Luzula* will be removed from the plants one month before sampling, so that the harvested material will contain the dead material which has accumulated within the preceding month. In months in which maximum or minimum values are expected, the number of parallel samples will be increased to twenty. Moreover, the plant density will be counted in ten $10 \times 10$ m sub-plots instead of one, in order to get an idea of the variance of changes in the number of plants.

## Reference

NEWBOULD, P. J.: Methods for estimating the primary production of forests. Oxford: Blackwell 1967.

# H. Chemical Composition of Plants of the Field Layer Preliminary Report

R. BORNKAMM and W. BENNERT

## I. Introduction

Production of plants is mostly evaluated by determining dry matter production. This measure indeed allows a comparison of different stands but fails to solve such special problems as estimating the influence of environmental factors on production or calculating cycles of certain elements or compounds. The determination of the chemical composition of herbage dry matter is expected to supply at least partial solutions to the above-mentioned problems.

For this reason we are investigating first the major classes of chemical compounds in plants constituting the herbage and moss layer in the beech forest (B 1) at different seasons. This project is not yet finished: so far, estimates are available of total nitrogen, protein, total lipids, and total ash contents on three sampling dates (13. 5. 1967, 15. 7. 1967, 18. 10. 1968). By combining data from biomass determination with those obtained by chemical analyses, it is possible to quote quantities per area for the chemical compounds investigated.

## II. Plant Composition

### 1. Methods

After harvesting samples were thoroughly washed, separated into above-ground and subterranean parts (leaves, stems and roots in the case of *Fagus silvatica* and *Sorbus aucuparia*), dried at 105° C, ground, and stored in powder form.

Total nitrogen was estimated as the sum of trichloroacetic acid (TCA)-soluble N and TCA-insoluble N, both determined by the micro-Kjeldahl method. Protein was calculated as TCA-insoluble nitrogen $\times 6.25$. For the estimation of total lipids, the methanol extract and subsequent ether extract were combined and the solvents were vaporized. The ether-soluble part of the residue was determined gravimetrically (RADUNZ, 1966, procedure slightly modified).

The total ash was estimated by weighing after combustion at 500° C till weight remained constant.

For details of nitrogen and ash determination, see: Association of Official Agricultural Chemists (1960); PIPER (1950). The values are given as percentage of dry matter.

### 2. Some Results and Discussion

The plants investigated can be divided into two groups. The first group comprises broad-leaved, deciduous, mesotrophic plants *(Fagus silvatica, Sorbus aucuparia, Poly-*

*gonatum verticillatum*, *Oxalis acetosella*, and *Dryopteris carthusiana*) with the following
chemical qualities: the nitrogen and protein contents and (usually) the total lipid
and ash contents are higher than in the second group. The N content of the above-
ground parts exceeds that of the subterranean ones. Moreover, the above-mentioned
plants exhibit distinct variations which generally followed the same trend in all
5 species (e. g. *Fagus silvatica* and *Oxalis acetosella* in Table 1).

Table 1. *Chemical composition of four species of the field layer at different seasons*
Abbreviations: *a* above ground parts; *s* subterranean parts

| | | | total nitrogen | TCA-sol. nitrogen | protein | total lipids | total ash |
|---|---|---|---|---|---|---|---|
| | | | | (in % of dry matter) | | | |
| A) *Fagus silvatica* L. (seedlings and young plants) | | | | | | | |
| | May 1967 | leaves | 0.83 | 0.08 | 4.69 | 2.1 | 4.5 |
| | | stems | 0.32 | 0.07 | 1.56 | 0.9 | 3.2 |
| | | roots | 0.47 | 0.07 | 2.50 | 1.4 | 6.6 |
| | July 1967 | leaves | 0.59 | 0.02 | 3.56 | 1.5 | 4.2 |
| | | stems | 0.23 | 0.02 | 1.31 | 1.1 | 2.9 |
| | | roots | 0.29 | 0.02 | 1.69 | 1.7 | 4.8 |
| | Oct. 1968 | leaves | 0.50 | 0.06 | 2.75 | 5.8 | 4.7 |
| | | stems | 0.32 | 0.10 | 1.38 | 2.1 | 2.7 |
| | | roots | 0.36 | 0.11 | 1.56 | 2.0 | 2.6 |
| B) *Oxalis acetosella* L. | | | | | | | |
| | May 1967 | a | 1.02 | 0.17 | 5.31 | 2.3 | 6.9 |
| | | s | 0.64 | 0.16 | 3.00 | 0.9 | 5.7 |
| | July 1967 | a | 0.77 | 0.09 | 4.25 | 2.3 | 5.2 |
| | | s | 0.52 | 0.09 | 2.69 | 1.3 | 5.0 |
| | Oct. 1968 | a | 0.77 | 0.18 | 3.69 | 8.3 | 6.2 |
| | | s | 0.60 | 0.30 | 1.88 | 3.3 | 3.8 |
| C) *Deschampsia flexuosa* (L.) Trin. | | | | | | | |
| | May 1967 | a | 0.48 | 0.07 | 2.56 | 0.9 | 7.0 |
| | | s | 0.42 | 0.05 | 2.31 | 1.1 | 1.9 |
| | July 1967 | a | 0.53 | 0.07 | 2.88 | 1.5 | 6.2 |
| | | s | 0.59 | 0.05 | 3.38 | 1.7 | 3.6 |
| | Oct. 1968 | a | 0.51 | 0.11 | 2.50 | 2.4 | 5.1 |
| | | s | 0.37 | 0.07 | 1.88 | 2.3 | 2.8 |
| D) *Luzula albida* (Hoffm.) DC. | | | | | | | |
| | May 1967 | a | 0.60 | 0.10 | 3.13 | 1.5 | 5.3 |
| | | s | 0.51 | 0.11 | 2.50 | 0.9 | 3.2 |
| | July 1967 | a | 0.56 | 0.05 | 3.19 | 1.3 | 8.6 |
| | | s | 0.50 | 0.07 | 2.69 | 1.2 | 4.0 |
| | Oct. 1968 | a | 0.51 | 0.10 | 2.56 | 4.1 | 5.2 |
| | | s | 0.41 | 0.14 | 1.69 | 2.0 | 4.1 |

The second group includes the grass-like or small-leaved, partly evergreen, more
oligotrophic plants (*Deschampsia flexuosa*, *Luzula albida*, *Juncus effusus*, *Carex pilulifera*,
and *Polytrichum attenuatum*). Their characteristics are: lower nitrogen, protein, and
lipid contents; less obvious seasonal variations of these chemical compounds than in
group 1 (e. g. *Deschampsia flexuosa* and *Luzula albida* in Table 2.)

The data of three samplings collected in two different years may be combined with reservations into one preliminary seasonal cycle (Table 1). Usually the total nitrogen and protein contents decrease from May to October (except in *Deschampsia*), while total lipids increase (except in the green parts of *Dryopteris*). The total ash content exhibits a great variation, which can only be explained by analyses of the single elements it contains. Extremely low values are found in the stems of young beech *(Fagus)* and rowan *(Sorbus)* plants.

In all cases, even in group 1, the N and protein contents of the Solling plants must be considered very low. Other authors have found in *Deschampsia flexuosa* nitrogen contents of 0.75% (ZARZYCKI, 1968), 1.38% (GRABHERR, 1942, cited after ELLEN-BERG, 1963), 1.7% (DUVIGNEAUD et DENAYER-DE SMET, 1963), 1.14−2.05% (HÖHNE, 1962, 1963). The same is true of the other species.

The investigations are being continued in order to determine other groups of compounds, such as soluble and polymeric carbohydrates, lignin, and C content.

# III. Biomass and Chemical Compounds per Area

## 1. Methods

In order to maintain our research area intact, we could not use a destructive harvesting method (NEWBOULD, 1967) inside the experimental plot. Therefore we counted the number of individuals of each species in our research area and collected a known number of individuals (or individual shoots) of about the same size outside the enclosure. A similar method has been described by BRECHTL and KUBICEK (1968). These data enable us to calculate biomass and chemical compounds per area for any species investigated.

## 2. Some Results and Discussion

Table 2 gives some of the results obtained in October 1968. The total biomass amounts to 8.04 g/m² (80.4 kg/ha), from which only 1.8 kg/ha is protein, 3.4 kg/ha ash, 2.5 kg/ha lipids.

Table 2. *Biomass and chemical compounds per area in the field layer of the beech forest (B 1) in October 1968*

|  | biomass (g/m²) | protein (mg/m²) | total ash (mg/m²) | total lipids (mg/m²) |
|---|---|---|---|---|
| *Polytrichum attenuatum* | 3.36 | 86.3 | 150.7 | 135.1 |
| *Luzula albida* | 2.39 | 49.4 | 109.5 | 69.6 |
| *Fagus silvatica* | 0.88 | 14.3 | 25.6 | 22.2 |
| *Polygonatum verticillatum* | 0.56 | 9.3 | 21.6 | 3.3 |
| *Deschampsia flexuosa* | 0.41 | 8.6 | 14.7 | 9.6 |
| *Juncus effusus* | 0.23 | 4.5 | 10.2 | 3.3 |
| *Dryopteris carthusiana* | 0.11 | 2.7 | 4.5 | 5.0 |
| *Oxalis acetosella* | 0.06 | 1.5 | 2.7 | 3.0 |
| *Sorbus aucuparia* | 0.04 | 0.6 | 1.2 | 0.8 |
| *Carex pilulifera* | 0.00 | 0.1 | 0.3 | 0.0 |
| total | 8.04 | 177.3 | 341.0 | 251.9 |

The figures for the biomass seem very low compared with data from Ellenberg, 1939 in Querco-Carpineta and other findings (e.g. Höhne, 1962; Rajchel, 1965; Andersson, 1970), whereas figures of Eber concerning the same beech forest area (B 1) are even lower. The method of Eber (see chapter G) is to be regarded as the more excact one, because it refers to a much larger research area. Together with Eber, we are planning to take monthly samples next summer for estimation of both plant productivity and production of chemical components; this should enable us to describe exactly changes in the chemical composition of each species and the whole field layer per unit area.

# References

Andersson, F.: Ecological studies in a Scanian woodland and meadow area, Southern Sweden. II. Plant biomass, primary production and turnover of organic matter. Botan. Notiser **123**, 8—51 (1970).

Association of Official Agricultural Chemists (AOAC): Official methods of analysis of the Association of Official Agricultural Chemists. Washington 1960.

Brechtl, J., Kubiček, F.: Beitrag zur Messung der primären Produktivität der Kräuterpflanzenschicht von Waldgesellschaften. Biologia **23**, 305—316 (1968).

Duvigneaud, P., Denayer-De Smet, S.: Distribution de certains éléments mineraux (K, Ca et N) dans les tapis végétaux naturels. Bull. Soc. Franc. Physiol. Vég. **8**, 1—8 (1963).

Ellenberg, H.: Über Zusammensetzung, Standort und Stoffproduktion bodenfeuchter Eichen- und Buchen-Mischwaldgesellschaften Nordwestdeutschlands. Mitt. Flor.-Soziol. Arbeitsgem. Niedersachsens **5**, 3—135 (1939).

— Vegetation Mitteleuropas mit den Alpen, 943 S. Stuttgart: Ulmer 1963.

Höhne, H.: Vergleichende Untersuchungen über Mineralstoff- und Stickstoffgehalt sowie Trockensubstanzproduktion von Waldbodenpflanzen. Arch. Forstwesen **11**, 1085—1141 (1962).

— Der Mineralstoff- und Stickstoffgehalt von Waldbodenpflanzen in Abhängigkeit vom Standort. Arch. Forstwesen **12**, 791—805 (1963).

Newbould, P. J.: Methods of estimating the primary production of forests. Oxford: Blackwell 1967.

Piper, C. S.: Soil and plant analysis. New York: Interscience Publishers 1950.

Radunz, A.: Chlorophyll- und Lipidgehalt der Blätter und Chloroplasten von *Antirrhinum majus* in Abhängigkeit von der Entwicklung. Z. Pflanzenphysiol. **54**, 395—406 (1966).

Rajchel, R.: Net primary productivity of the herb layer in two forest associations of the Ojców National Park (Southern Poland). Fragm. Flor. et Geobot. **11**, 121—150 (1965).

Zarzycki, K.: Experimental investigation of competition between forest herbs. Acta Soc. Botan. Polon. **37**, 393—411 (1968).

# I. Primary Production of a Meadow (Trisetetum flavescentis hercynicum) with Different Fertilizer Treatments — Preliminary Report

B. Speidel and A. Weiss

## I. Introduction

The aim of the investigations is to determine the total production by the organs above and below ground, and to observe their seasonal development.

In order to be able to evaluate both the present state of the yield and the potential yield, a fertilizer experiment with the following 3 variants was laid out in the autumn of 1966:

a) unfertilized; 2 harvests
b) 90 kg/ha $P_2O_5$ + 120 kg/ha $K_2O$; 2 harvests
c) 120 kg/ha $P_2O_5$ + 240 kg/ha $K_2O$ + 200 kg/ha N; 3 harvests
   (portions of N-fertilizer: 90 + 60 + 50).
   (six replications of each variant.)

The initial vegetation corresponded, in the phytosociological sense, with the Red Fescue facies of the Golden Oat meadow (*Festuca rubra*-facies of the *Trisetetum flavescentis*, chapter Z).

## II. Methods

### 1. Determination of the Changes in the Composition of Species Caused by Fertilizing

Twice a year (at the end of May before the first harvest, and at the end of August before the second harvest) by estimating the proportions of the individual species before cutting (according to Klapp, 1949).

### 2. Measurement of the Agriculturally Utilizable Production

The unfertilized and the PK plots are mown twice (around the middle of June and the end of August), and the NPK plots three times (end of May, beginning of July and end of August). The last harvest thus takes place on all fertilizer variants at the same time, in order to create the same conditions for wintering the plant stock. The mowing is carried out with a motor mower set at the cutting height of 4 cm. After the fresh weight has been taken, a 1 kg sample is taken from each plot, in order to determine the dry weight and the constituent substances.

### 3. Investigation of the Dynamics of Increment and Losses

For this purpose the total matter above and below ground is measured on four replicates of each fertilizer variant on nine different dates:

a) before the start of the vegetation period (around the end of March)
b) approx. 4 weeks after the start of the vegetation period (around the end of April)
c) at the first harvest of the NPK variant (around the end of May)
d) at the first harvest of the unfertilized and the PK variants (around the middle of June)
e) at the second harvest of the NPK variant (around the beginning of July)
f) at the last harvest of all the variants (around the end of August)
g) 4 weeks after the last harvest (around the end of September)
h) 8 weeks after the last harvest (around the end of October)
i) at the end of the vegetation period (around the end of November)

## 4. Above-Ground Matter

Per plot 3 areas $25 \times 25$ cm are sampled. So, with four replicates, 12 separate measurements for each fertilizer variant are necessary:

a) above-ground matter cut off to leave 4 cm stubble
b) stubble cut off at surface of the ground and separated into living and dead matter.
c) collection of matter that has fallen to the ground as litter (this is later put together with the dead matter from the stubble).

## 5. Underground Matter

Per plot 4 root-borings (so with 4 replicates, 16 separate measurements for each fertilizer variant). Boring depth at least 40 cm, if possible as much as 60 cm. Diameter of the borer 7 cm (Borer according to Vetter, 1964). The core extracted is in each case cut up into five sections: 0 to 5 cm, 5 to 10 cm, 10 to 20 cm, 20 to 40 cm and 40 to 60 cm of the boring depth. Finally, in order to determine the absolute dry weight of the roots, each of these sections is softened in water and then washed out with a fine spray over three different sieve meshes (the finest mesh being 1.5 mm).

## 6. Determination of the Constituent Substances

a) In order to calculate the organic matter and the starch units of the agriculturally utilizable production, the content of raw fibre, protein and ash, as well as the mineral elements, is analysed (for method, see Nehring, 1960).

b) The amount of organic matter and reserve carbohydrates in the stubble and roots is measured (organic matter: the difference after incineration; carbohydrates: colorimetric analysis after hydrolysis; method, according to Dubois (see KakÀč and Vejdelek, 1966).

## 7. Investigation of the Root Activity

For this purpose 4 root boxes were installed which allow observation of the root growth to a depth of about 40 cm (2 boxes on unfertilized plots, 2 on NPK fertilized plots). During the frost-free period, from around mid-March to the beginning of December, the roots are observed at two-weekly intervals; the number of roots which show an increase in length since the previous observation as well as the number of new roots (active roots) are counted; furthermore the number of roots with no measurable increase in length (inactive roots) is determined, and the number of those which have clearly died recently, as recognized by their brown and shrivelled tips.

In addition the increase in length of the active roots between each observation is measured in mm.

## 8. Observation of the Root Systems

By the use of a nail-board (see SCHUURMAN and GOEDEWAAGEN, 1965) the depth and extension of the roots are examined once a year, in all cases in the autumn during the period of greatest root activity.

# III. Some Results

## 1. Changes in the Species Composition Caused by Fertilizing

Clear differences in the composition of the species, corresponding to the use of the individual fertilizer variants, were already evident in the first year of the experiment; these differences increased in the succeeding years. The most striking observation was the known advancement of the legumes caused by PK fertilizer, and their almost complete suppression by N fertilizer, accompanied by a corresponding advancement of the grasses. On the unfertilized plots, moreover, there was a yearly increase in *Anthoxanthum odoratum* (at the first harvest) and above all of *Agrostis tenuis* (at the second harvest), which indicates a deficiency of nutrients in the soil. On the fully fertilized plots, on the other hand, both these species are found only in small numbers. and on the PK plots they are found in quantities which lie somewhere between these two extremes.

## 2. The Utilizable Production

Since the experiment was started, the following yields have been recorded (Table 1).

Table 1. *Yields of unfertilized and fertilized plots* (dry matter, t/ha)

|  | unfert. | PK | NPK |
|---|---|---|---|
| 1967 1st exp. year | 4.42 | 5.36 | 8.27 |
| 1968 2nd exp. year | 2.39 | 4.22 | 7.84 |
| 1969 3rd exp. year | 2.10 | 4.94 | 6.63 |
| average for 1968/69 | 2.25 | 4.58 | 7.24 |

Since the previous history of the experimental area is not known, and moreover since the species composition was in a state of rapid change during the first experimental year, it is better not to take yields for 1967 into consideration. From the results for 1968 and 1969, it is seen that the yield with PK fertilizer is twice as high as the yield without fertilizer, and that the yield with NPK fertilizer is more than three times as high as that without fertilizer. The result also shows that a *Trisetetum* that is poor in upper grass but rich in *Festuca rubra* can, when well fertilized, as a consequence of its dense turf, get fairly close to the potential yield of an *Alopecurus-Arrhenatheretum*, which is rich in upper grass always somewhat loosely turfed.

## 3. The Dynamics of Increment and Losses

### a) Above-Ground Matter

The different growth rates of the three fertilizer variants are clearly seen in the time required for the growth of the living matter above ground. While in both years of observation (1968 and 1969) the NPK plots already showed a total of 3 t dry matter/ha at the end of May, this quantity was not reached by the PK plots until three weeks later (mid-June), and was not reached by the unfertilized plots until as much as 3 or respectively 2 months later. Further production of utilizable matter above ground continues until about mid-September, when vegetative growth more or less ceases, or is translocated below ground.

The amount of living organic matter remaining after the agriculturally utilizable production has been cut (stubble) varies during the course of the year, uniformly for all fertilizer variants, by a quantity in each case of about 1.0 t/ha dry matter.

Similarly, in the dead matter above ground (litter), as measured on the different dates for observation, there were small variations but no significant differences between the individual fertilizer variants. The decisive factor affecting the absolute amount here, however, seems to be the seasonal weather conditions. In both years of observation (1968 and 1969) about 1.5 t/ha dry matter was measured in mid-April. With the beginning of the warm weather a very rapid decomposition seems to follow, so that at the end of May (in both years) only about 0.5 t/ha dry matter was found. In 1968 this quantity remained more or less constant until the middle of October, and so according to this a balance had developed between the decomposition and the further accumulation of dead matter. Not until November (with the beginning of the cold season and the associated decline in biological activity) was there an increase in the quantity of crop-litter to about 1.5 t/ha dry matter. In 1969, on the other hand, there was already a rise at the end of June to about 1.0 t/ha dry matter, and as early as mid-August the quantity of litter was a little over 1.5 t/ha dry matter. The reason for this is most probably to be found in the long summer drought.

### b) Organic Matter Below Ground

In all 3 fertilizer variants the largest quantity of root biomass was recorded in the spring (March) and in late autumn (October/November). On the other hand during the summer, i.e. during the period of growth above ground, there was a marked reduction in the quantity of root biomass. The reason for this probably lies partly in the more rapid turnover during the summer and partly in the fact that not only are less reserve substances stored during the period of growth, but also that those present in the spring are used chiefly for the production of above-ground matter. In accordance with this, is the finding that the unfertilized variant, that is the one with the least above-ground production, also has the smallest decrease in root biomass during the summer; the NPK variant, on the other hand, has the largest, while the PK variant again occupies a mid-way position (Table 2).

The difference between the respective maxima and minima represents the annual minimum production of root matter that is directly measurable. However, if the continuous but not directly measurable decomposition of dead roots (and at least

Table 2. *Ash-free root dry matter in t/ha 1968*

|           | unfert. | PK   | NPK  |
| --------- | ------- | ---- | ---- |
| Maximum   | 8.32    | 8.93 | 8.37 |
| Minimum   | 6.29    | 6.07 | 5.16 |
| Difference| 2.03    | 2.86 | 3.21 |

partial replacement by development of new ones) is taken into consideration, then the actual annual production could well be significantly higher.

On average, through the year, approximately 75% of all the root biomass is to be found in the uppermost 5 cm of the soil. The section from 5—10 cm contains about 13%. About 7% is found in the section from 10—20 cm, and about 4% in the section from 20—40 cm. The remaining 1% is in the section from 40—60 cm.

## 4. The Constituent Substances

The results of the chemical analysis of the harvest generally lie within the usual range for meadow hay. A detailed discussion of this will follow after the termination of the project, when the results for several years are available.

For the period from November 1967 to November 1968, we have details of the reserve substances contained in the roots for ten different dates. The total quantity of soluble carbohydrates (determined by hydrolysis) was greatest in the two November months (i. e. during the period of no vegetative growth) with approximately 13% (relative to ash-free root dry matter). During the winter there seems to be only a comparatively slow decrease, for at the end of March there was still about 8%. But one a month later (end of April) the contents had fallen to 2 to 3%. After this there was a slow build-up again, so that by the middle of June the level had risen to approximately 9%. It remained at this level until about the end of September. Then followed a further big increase, so that by the end of October the level had already reached approximately 14%, and this was still the case at the end of November. The results show that in the spring, during the first flush of growth, very many reserve substances are mobilized and are required for the growth of above-ground matter. The deficit thus caused is partly made up again during the ensuing period of high vegetative growth (until about mid-June) when the meadow plants show their greatest above-ground development, and as a consequence probably produce a certain surplus of assimilates. It seems, however, that during the ensuing late summer these newly formed assimilates are required and used for further growth of the above-ground matter. Not until the autumn, when there is no more significant above-ground growth and yet the plants are still fully capable of assimilation, are the required reserve substances built up.

## 5. Root Activity

In accordance with the biological growth rhythm of the meadow plants, the number of active root tips present at any one time during the vegetation period shows three principal maxima: one in the spring or early summer (April to May) during the first period of intensive above-ground growth, and a second in the summer (around August). Between these two a further, but smaller and indistinct, maximum may be observed in June/July. The third maximum appears in the autumn (October/ November) and is in all probability connected with the build-up of reserve substances, as at this time both the absolute root biomass and the soluble carbohydrates it contains are on the increase, while, on the other hand, there is no apparent above-ground growth. The first maximum, in particular, seems to be very dependent on the weather. In 1969, for example, after the relatively cold spring, it appeared two or three weeks later than in 1968. Quite apart from this, the time of the appearance of the maxima is also dependent upon the fertilizer treatment. On the NPK variants they always appear at least one observation interval earlier than on the unfertilized variants.

The daily increment of the active roots is on the average slightly higher on the unfertilized variants than on the NPK variants. Since, however, the absolute number of the active roots on the NPK plots is significantly higher than on the unfertilized plots, there is, nevertheless, a distinctly higher absolute total increment per day. The respective numbers of inactive roots show variations similar to those of the active roots, but with the difference that the times when the maxima appear are delayed by a few weeks according to the period of activity of the root tips.

## IV. Further Investigations Proposed

In order to support and supplement the results obtained so far, and more particularly because of the marked annual variations in the climate, the tasks commenced are to be continued for another year.

The observations of root activity can only be carried out at relatively long intervals (every two weeks) because of the distance of the experimental field from the institute. In order to shorten the intervals and thus gain a more exact picture of the progress of the activity curves, further root-boxes, which are observed twice a week, were installed during the summer 1969 on the institute's experimental field in a meadow of similar botanical composition to that of the experimental field in the Solling. In this way it should be possible in future to determine the effects of the ecological factors on the root growth, and thus also to obtain valuable information that will assist the evaluation of the data and results from the experimental field in the Solling.

The question arises in this context, to what degree the data gained from the root-boxes are representative for the plant community itself. To settle this question, random samples are to be taken of soil cores from each of the various depth sections, and the number of active root tips in them counted; these results will then be compared with those obtained from the root-boxes.

To supplement the root-borings carried out on the experimental meadow in the Solling, it is also intended to make random borings in the neighbourhood of the experimental field on different plant communities. This will enable a comparison to be made of the production conditions of different grassland communities in the same altitudinal zone.

# References

KAKÁČ, B., VEJDĚLEK, Z. J.: Handbuch der Kolorimetrie. Vol. III, Jena: G. Fischer 1966.

KLAPP, E.: Landwirtschaftliche Anwendungen der Pflanzensoziologie. Stuttgart: Ulmer 1949.

NEHRING, K.: Agrikulturchemische Untersuchungsmethoden für Dünge- und Futtermittel, Böden und Milch. 3. Aufl., Hamburg: Parey 1960.

SCHUURMAN, J. J., GOEDEVAGEN, M. A. J.: Methods for the examination of root systems and roots. Wageningen: Agr. Publ. Doc. 1965.

VETTER, H., SCHARAFAT, S.: Z. Acker- u. Pflanzenbau **120**, 1964.

# J. Green Area Indices of Grassland Communities and Agricultural Crops under Different Fertilizing Conditions

E. Geyger

## I. Introduction

The primary production of green plants is closely related to the sum total of assimilatory plant surfaces. Among grasses and herbaceous plants it is not only the flat, spreading leaves which make up these surfaces, but also three-dimensional organs such as rolled leaves, leaf sheaths, stems, and green buds or fruits. The estimation of the assimilatory value of these rounded plant parts is not to be discussed here; it is surely different from case to case. A direct determination of this value for the species discussed in this paper would be welcome in order to allow a complete interpretation of the data collected. Taken as a whole, the sum of green plant surfaces is usually a multiple of the ground surface it covers, i.e. the "green area index" (G.A.I.) is greater than 1. For the plant communities which I investigated, the face-up surface of the flatly spread out green leaves, and the projection area (shadow area) of the three-dimensional green plant parts were included. For this reason, I modified the term "leaf area index" (L.A.I.), normally used in reference to woods, accordingly.

## II. Methods

In order to assess the variously shaped surfaces of herbaceous plant communities, a uniform measuring technique was developed (GEYGER, 1964), of which only the most important points are summarized here:

After removal, the *leaves* are sorted according to species or shape and spread out evenly on a white, rectangular plate of known size. The degree of coverage of the plate by green leaves can be sufficiently exactly determined with the help of home-made graduated estimation scales. This makes it unnecessary to measure each individual leaf.

With the help of specially constructed tables, the projection area of *convex plant parts* (e. g. round, three- and four-edged stems) is directly determined after measuring the diameters by the use of a binocular microscope or a calliper.

Samples must be taken in a manner that enables their relation to the ground area to be established. In order to do this, representative weight samples from the harvest of agricultural experimental plots are taken and their dry weight is measured. The ground surface area per sample is determined from the ratio of the dry weights of the samples to the yield per hectare. If one wishes to work with samples taken between the harvest dates of the experiments (e. g. for growth curves), or with plots for which no yield per ha is determined, the samples must be taken from a measured ground area. Here, too, the choice of representative samples is important. Exactness can be enhanced if ground-area-related samples and weight-related harvest samples are evaluated together. The natural variability of the density of plant growth and of species distribution in grassland communities makes it necessary to take a large number of parallel samples. Therefore I tried to develop a rational estimation process for determining the green surface area. The limitations of the exactness of my methodology were thoroughly discussed in a previous paper (GEYGER, 1964).

## III. Some Results

### 1. Grassland Communities

The green area indices (G.A.I.) of the variously fertilized plots in the grassland community fertilization experiment (see SPEIDEL and WEISS, chapter I) differed according to fertilizer rate. These values also showed great differences in the years 1967 and 1968: The expected increase in fertilizer effect in the second year was cancelled out by the negative effect of unfavorable weather conditions. Thus, the NPK plots, for example, achieved indices of 7.5 and 7.4 for the first two mowings in 1967, and approximately 5.8 for the third mowing. In 1968, the indices were approximately 4.0 for the first and about 5.5 for the following two mowings. In the unfertilized plots, soil impoverishment and unfavorable weather worked together in 1968 so that the index fell from about 3.8 in 1967 to about 1.9 in 1968. The indices of the PK plots lay between these, and were likewise lower in 1968 than in 1967.

The effect of soil impoverishment was evident in the decreasing plant surface development of samples from 0-plots outside the fertilization experiment which had previously been cultivated in the usual manner (some mineral fertilizer; grazing by cattle after the second mowing). Here the 1968 indices also fell to less than half the 1967 ones.

For the purposes of comparison, the positive effect of multiple mowings on the total production of green plant surfaces was checked on plots which were mown only once (in September). Here, despite the much longer growth period, the indices of the NPK plots, in particular, were scarcely higher than those from one mowing of the thrice-mown plots. The upper limit was reached rather quickly and then, with continued increase of the plant mass above ground, the green surface area decreased again. In contrast to this, the unfertilized plots reached their maximum green area in July.

### 2. Agricultural Crops

The agricultural experiment (Prof. BAEUMER, Institut für Pflanzenbau, University of Göttingen) included, in 1967, three variously fertilized maize plots and, on the edges of these plots, three identically fertilized field-weed communities with suppressed maize plants. The weed communities showed relatively small differences due to fertilizing; their G.A.I. lay between 4 and 5. The "yields" were 3,500—4,500 kg/ha dry matter.

The maize experiment brought the surprising result that high yields (14,000, 10,000 and 6,000 kg/ha, respectively) were accompanied by relatively low surface development. With the inclusion of stems and green husks, the G.A.I. was 5.2 under maximal fertilizing conditions, 3.4 under intermediate fertilizing conditions, and about 1.3 for the unfertilized plots.

In 1968, a fertilization experiment with *Lolium multiflorum* was begun (Baeumer). The determination of the G.A.I. yielded the following values: after approximately eight weeks' growth, from June to August 1968, the index was 11.6 and after another eight weeks from July to September on other plots 9.3. The corresponding yields were 5,500—6,000 kg/ha and 3,500—4,000 kg/ha.

## IV. Weight/Surface Ratio

From the green area indices ascertained, an informative value can easily be calculated: the relation of the sum total of green surfaces to total above-ground plant biomass at the time the sample was taken. If only the plant substance which is still green is used in the calculation, indications are obtained as to the degree of hygromorphism or xeromorphism of the plant community; if the calculation is based on the total amount of above-ground substance produced until then, the ratio will be modified by the state of maturity of individual plants, as well as by the phenomenon of yellowing, caused by shade. I shall now give some examples of ratios in milligrams total dry weight to square centimeters of green surface. The lowest values, between 3.3 and 4 mg/cm$^2$, were obtained in 1967 for the NPK-fertilized meadow plots. The values varied from 4—5 mg/cm$^2$ on the NPK plots in 1968, and on the PK meadow plots in 1967 and 1968, and for *Zea mays* as well as for *Lolium* (in each case after eight weeks of growth).

The unfertilized meadow plots in 1967 and 1968, as well as the unmown PK plots and the unfertilized meadow plots in 1967, gave weight/surface ratios from 5—7 mg/cm$^2$. Values up to 11 mg/cm$^2$ were reached by the unmown NPK plots in 1967, by all unmown plots in 1968, as well as by the field-weed communities at all fertilization levels.

For maize, the weight/surface ratio rose, during the growth and maturation periods, from 4.0—4.5 mg/cm$^2$ in July, to 33 mg/cm$^2$ in October.

## V. Comparison and Discussion of the Reported Data

The values quoted in this communication as examples of the green area indices of the variously fertilized plant communities, the dry matter produced above and below ground level, and the relationship of weight to surface of green parts, allow the recognition of several tendencies and relationships:

1. Fertilizing increased the production of green surfaces in the NPK meadow plots up to 4 times that of the unfertilized plots, and reduced by 50% the time required to achieve maximum green surface production. The higher indices of *Lolium multiflorum* in the agricultural experiment, then, may be due to higher fertilizer rates, as well as to the plant variety.

2. The weather conditions in the second year of experimentation had a negatively interacting influence with the effect of the fertilizer. On the fertilized plots, green area indices and yields decreased. On the unfertilized plots, the negative effects of the lack of fertilizer were supplemented and emphasized by unfavorable weather conditions.

3. Fertilizer, and above all nitrogen fertilizer, considerably increased the degree of hygromorphism of the plant communities. This increased the green area

indices relatively more than the yields, e. g. in the NPK meadow plots, three times more surface with a doubling of the yield as compared with unfertilized plots.

4. In the grassland community, the sum total of green surface which was increased by fertilization was not paralleled by a higher production of underground matter; on the contrary, this production was, on the average, approximately 15% higher on the PK and unfertilized plots (SPEIDEL and WEISS, chapter I). On all levels of fertilization, the periods of greatest above-ground development were periods of reduced root mass production.

For maize, high yield, low production of underground matter, and average green surface production were linked together. The total yield was much higher than for the meadow. Besides the higher fertilizer rate, this could be connected with the maize plant's economy of respiration.

5. In conclusion, the green area indices of differently structured plant communities in similar habitats are compared (see Table 1), in each case for one vegetation period during which the plants were allowed to grow continually (i.e. without mowing or cutting).

Table 1. *Green area indices (G.A.I.)*

| Community | treatment | 1967 | 1968 |
|---|---|---|---|
| unmown meadow plots | unfertilized | 5.7 | 4.8 |
| | PK | 6.7 | 5.4 |
| | NPK | 8.0 | 6.6 |
| field-weed community | unfertilized | 4.5 | — |
| | normally fertilized | 4.4 | — |
| | maximally fertilized | 5.0 | — |
| beech forest (B 1) | unfertilized | 6.4 | 5.7 |

The leaf area indices of the beech forest were determined by HELLER (chapter B). There exists so little ground vegetation, that its surface area may be neglected.

Worthy of mention is that, in comparing herbaceous communities with the beech forest, the forest is more readily associated with fertilized than with unfertilized habitats. The application of fertilizers obviously replenishes the agricultural ecosystems with the mineral substances removed by harvesting, while in the forest ecosystem these substances are retained in the form of litter.

All G.A.I. values in Table 1 lie in the intermediate range. The improvement of habitat due to the application of fertilizer is not as pronounced when a single mowing is made at the end of the vegetation period as when several mowings are made during that period. This is due to the fact that the upper limit for the production of green surfaces is determined by light: the plant stand becomes so dense that, for lack of light, photosynthesis is no longer possible near ground level, and once-green parts of the plants turn yellow.

# Reference

GEYGER, E.: Methodische Untersuchungen zur Erfassung der assimilierenden Gesamtoberflächen von Wiesen. Ber. Geobot. Inst. ETH, Stftg. Rübel, Zürich 35, 41—112 (1964).

# K. Methodological Studies to Distinguish Functional from Non-functional Roots of Grassland Plants

Ch. Sator and D. Bommer

## I. Introduction

Studies of the primary production in terrestrial plant communities are not fully comprehensive without records of the total weight of the root system. But root determinations lack the precision of measurements of the shoot system, especially in perennial plants, because of the difficulty of distinguishing between functional and non-functional roots. So far no adequate methods are available to decide whether a root is alive or not. In methodological studies started two years ago, we selected three approaches for developing a reliable method for use in ecological programs.

## II. Material and Methods

### 1. Partial Chemical Characterisation of Cell-Wall Substances

From pot experiments with *Trisetum flavescens* in the greenhouse, roots were sampled for analysis of different growth stages 8, 12, 16 and 20 weeks after germination. At the same time roots were killed by cutting off the green plant parts. Rotted root material was investigated after 4, 8, 12, 16 and 20 weeks. The samples of freeze-dried powdered root material were fractionated by a modification of the methods used by DEVER et al. (1968) and MAEDER (1960):

ether extraction 36 h
↓
cold water extraction 4 × 30 min
↓
ammonium oxalate extraction 2 × 24 h → pectin fraction
0.5% 90° C
↓
sodium hydroxide extraction 2 × 24 h → hemicellulose fraction
17.5% room temp.
↓
The residue was washed with ethanol, acetone,
and ether and suspended in 72% $H_2SO_4$ for 2 h
at room temp. with occasional stirring
and hydrolysed in 3% $H_2SO_4$ for 4 h at 105° C → cellulose fraction
↓
the non-hydrolysable residue was washed
with water, dried and ashed → lignin fraction

## 2. Quantitative Determination of the Dehydrogenase Activity of Roots at Different Stages of Growth

Plants of *Trisetum flavescens* from the same experiments as under 1. and in addition plants of a dicotyledonous species, were grown in the greenhouse and root samples taken at different stages of growth. For the determination of dehydrogenase activity, a modification of the method described by JAMBOR (1960) was used.

A sample of fresh roots was incubated in a buffered TTC-solution (pH 7.4) under vacuum, the colored material was filtered from the solution and the triphenyl-formazane extracted from the roots by acetone. The triphenylformazane solution was measured in a Beckman spectrophotometer DB-G at 490 nm.

## 3. Labelling with Carbon-14

Using a modification of a method described by UENO et al. (1967), carbon-14 was applied to the leaves of *Lolium perenne* * in a bottomless airtight plastic chamber in the field. Labelled carbon dioxide was produced in a generator connected to the assimilation chamber and two air pumps. The gas circulated through the chamber at a rate of about 35 l/h. By means of an iron trough (30 × 60 cm) filled with acid water, the plastic chamber (30 × 60 × 50 cm) was made airtight to the soil. Excess $^{14}CO_2$ was absorbed by an alkali solution using an air pump with a velocity of about 25 l/min. Root samples were taken by means of an auger and the roots washed free from soil in running water and freeze-dried overnight. Autoradiographs were developed after several days of exposure to X-ray film. The percentage of living root material in a sample was calculated by comparing the measured lengths of total and labelled roots.

## III. Some Results and Discussion

The first approach was chosen because lignin-like substances are decomposed very slowly, and pectins are easy soluble substances from the chemical point of view. So it was assumed that pectin would be decomposed easily by soil microorganisms. However, the results showed that, unless further characterization is made of the individual fractions, it cannot be definitively stated that pectin is decomposed very easily. Therefore no relation was found between the pectin and the lignin content suitable for the characterization of living or dead root material, and it seems doubtful whether any better results would be obtained from the relation of cellulose to lignin.

The second approach was also unsatisfactory because no clear quantitative relation could be found between living roots of different ages and dehydrogenase activity, although dead roots will not reduce TTC at all. Young root tips certainly have the most active dehydrogenase system and thick roots do not reduce as much TTC as thin roots, relative to a standard weight of the sample. But since it is too laborious to pick out the older and younger roots from a sample and to distinguish between thin and thick root parts, the TTC method does not seem to be useful, either.

---

* *Lolium perenne* was chosen as the single pure grass stand available in the field during this year close to the experimental station.

The third approach was based on the translocation of assimilates as cell meta-
bolites to the living parts of the root system. There may be some slight translocation
to dead parts by diffusion but this can be neglected because the dead roots do not
produce an image on the X-ray film. Therefore the use of carbon-14 proved to be a
good method for distinguishing between living and non-living root tissue. The
usefulness of the method at different growth stages, the amount of $CO_2$ needed, the
minimum duration of the treatment and various of ways recording the results are
being examined.

This method can be used as a field method, if the above-ground plant parts under
the chamber are removed after the treatment.

# References

DEVER, J. E., Jr., et al.: Partial chemical characterization of corn root cell walls. Plant
    Physiology **43**, 50—56 (1968).
JAMBOR, B.: Tetrazoliumsalze in der Biologie. Jena: Gustav Fischer 1960.
MAEDER, H.: Pflanzen-physiologische Untersuchungen an Rottestroh. Diss. Gießen 1960.
UENO, M. et al.: Living root system distinguished by the use of carbon-14. Nature **213**,
    530—532 (1967).

# L. Determination of Energy Values

M. RUNGE

## I. Introduction

The characteristic function of primary producers consists in transforming radiant energy to chemical energy. Part of the gross production, i.e. the total energy converted into organic compounds, is given off again through respiration of the primary producers. The remainder, the net production, is successively converted into heat via the consumers, i.e. animals and heterotrophic plants. In ecosystems, then, conversion of materials is synonymous with energy conversion. For the purpose of comparing the intensities of production and conversion of different ecosystems, or components of such systems, energy data are more suitable than dry matter data, since the energy content of the various plant parts and animal organs varies considerably.

The Solling Project presented an opportunity to compare energy conversions in different types of ecosystems. Therefore, simultaneously the determination of the amount and production of biomass in those ecosystems, the energy content (cal/g = energy value) of the quantitatively most significant components (e. g. boles, branches, leaves, etc.) was determined. A detailed report is presently being prepared.

Here, the amplitude of variation in the energy values found is presented as a supplement to the reports on ascertaining the production of the various vegetation types investigated. With regard to simplifying subsequent studies, such as the planned "Minimal Programs", the following questions are of special interest:

1. How great are the differences in energy values between the vegetation types (beech and spruce forest, meadow, field) and between their most important components?

2. How great are variations of energy values within the individual components?

3. How great are the seasonal variations in energy values?

This information determines the number of analyses necessary to insure sufficient accuracy in determining the energy conversions of plant communities, and indicates the comparability of the findings of different authors.

## II. Methods

### 1. Preparation of the Samples

Determination of energy values was carried out essentially as described by LIETH (1968). A substantial part of the material investigated was made available by the research groups which record the amount and production of biomass at the various sites. Informations on the gathering of material may be found in the relevant reports (chapters E and I). Since the samples were dried at different temperatures after gathering, the initial step in the treatment of the samples was a uniform drying at 105° C. Comparative studies revealed that, after drying at 105° C, the energy value was usually slightly higher than after drying at 80° C. Nevertheless, it can be assumed that in some cases volatile components are lost by drying at either 105° or 80° C. The significance of these losses in determining the energy value is presently being investigated in a series of experiments with various drying procedures.

Fresh material collected for special investigations was likewise dried at 105° C in a circulation oven. After being dried, all grindable material was ground in a disc-type swing mill (Siebtechnik GmbH), and then pressed into pellets by means of a pressing set, as described by Lieth (1968), and a hydraulic press. The weight of the pellets generally lay between 0.9 and 1.3 g, which resulted in a rise in temperature of 2—3° C after combustion. Material not suitable for pressing into compact pellets, such as bark and buds, was incinerated in acetobutyrate capsules. After being compressed, the pellets were again dried at 105° C and then stored over silica gel in an exsiccator. Ungrindable material (wood, bark) was cut to appropriate size (pieces of approximately 1 g) and then treated like the pellets.

## 2. Determination of Energy Value

Energy values were determined with an adiabatic calorimeter (Janke and Kunke KG). The energy equivalent (Roth and Becker, 1956) of the instrument was determined with benzoic acid and was checked at regular intervals.

The samples (pellets, pieces of wood, or capsules) were weighed to $^1/_{10}$ mg and then fastened with an ignition wire between the electrodes of the calorimeter bomb. The sample was enclosed in a quartz dish of known weight. Before closing the bomb, 5 ml distilled water were given on the bottom. Combustion took place under an oxygen pressure of 30 atm. Special care was taken to allow as little time as possible to elapse between removal of the sample from the exsiccator and ignition. Ignition was executed at the same temperature (deviations not greater than $\pm 0.1°$ C).

When the rise in temperature of the calorimeter after combustion had been ascertained, the quartz dish and the unincinerated remnants of the ignition wire were weighed separately. Traces of melted ignition wire which had dropped onto the dish were removed beforehand. Reweighing the quartz dishes served to indicate the ash residue of the pellets. Ignition wire remnants, if not detected, influence the measurement only insignificantly. As a control, however, the amount of ash residue was rechecked by means of incineration in an assay furnace (600° C).

When the weight of the sample and the rise in temperature after combustion, as well as the energy equivalent of the calorimeter are known, the energy value of the material can be calculated. A correction must be made to compensate for the partly burned ignition wire.

To make the analyses more rapidly, it proved advantageous to work with 2 combustion bombs and several quartz dishes. Thus, one bomb could be recharged while the other was in the calorimeter. After each individual analysis, the quartz dishes were removed from the bomb and returned to the exsiccator for storage. The reweighing to ascertain the amount of ash residue could then be executed at a later time, so that the time between analyses did not have to be used for this purpose.

The energy value can refer to either dry matter or ash-free matter. While the former value is generally used in calculating biomass and production, the latter allows differences in the chemical composition of different materials to be more easily recognized. The following discussion of the results concerns only the former value.

## III. Some Results

### 1. Energy Values of Different Plants

Generally, the spruce *(Picea abies)* material had the highest energy values. The energy values of the beech *(Fagus silvatica)* material lay somewhat below these, while the values of the hay from the meadow *(Trisetetum)* and from the *Lolium multi- florum* field were substantially lower. The herbaceous vegetation of the forest stands had somewhat higher energy values than the grassland vegetation, but was quanti- tatively of no importance for the ecosystem (Table 1).

Table 1. *Energy values of some of the quantitatively important components of the different vegetation types (cal/g)*

(The number of samples -n- is not equal to the number of analyses. On average, 4 analyses have been made per sample; s = standard deviation; s% = coefficient of variation).

| | $\bar{x}$ | $s$ | $s\%$ | $n$ | |
|---|---|---|---|---|---|
| *Fagus silvatica:* Boles and branches (7—35 cm ⌀) | 4694 | 18 | 0.4 | 18 | |
| Roots (0.5—7 cm ⌀) | 4782 | 25 | 0.5 | 14 | |
| Sun leaves | 5046 | 24 | 0.5 | 8 | } 18. 9. 1969 |
| Shade leaves | 4952 | 42 | 0.8 | 8 | |
| *Picea abies:* Boles and branches | 4846 | 41 | 0.8 | 15 | |
| needles | 4928 | 62 | 1.3 | 21 | |
| *Trisetetum* hay: NPK-treatment | 4315 | 137 | 3.2 | 6 | |
| PK-treatment | 4326 | 124 | 2.9 | 6 | } 3. 9. 1968 |
| O-treatment | 4407 | 50 | 1.1 | 6 | |
| *Lolium multiflorum* hay: treatment 1 | 4398 | 23 | 0.5 | 8 | |
| treatment 2 | 4366 | 45 | 1.0 | 7 | |
| treatment 3 | 4230 | 49 | 1.2 | 8 | } 7. 8. 1968 |
| treatment 4 | 4241 | 35 | 0.8 | 8 | |

This distribution of the energy values is consistent with the findings of other authors: material from herbaceous plants usually has the lower, and material from woody plants the higher energy values (GOLLEY, 1961; BLISS, 1962). OVINGTON and HEITKAMP (1960) found that the energy values of conifers were higher than those of deciduous trees.

The standard deviation of the mean values and the variation coefficients also indicate some characteristic differences. The highest standard deviations were recor- ded for the fertilized plots of the meadow. This finding is convincing, since the materi- al concerned was composed of a large number of species, whereas in the *Lolium multiflorum* plots, for example, only one species was represented. The standard deviation for the unfertilized treatment was not higher than those of the *Lolium multiflorum* plots, but it appears that fertilization promotes non-homogeniety, at least in the first years after the introduction of heavy fertilization. On the average, the analyses performed on hay from all harvesting dates showed the smallest standard deviation for the unfertilized treatment, and the greatest for the NPK treatment. In this respect the graduation in the example presented (Table 1) is typical.

The variation coefficients of the other materials investigated were consistently under 1%, indicating that the samples are very homogeneous in respect to energy value. At the same time, these small deviations lead to the conclusion that the accuracy of energy value analyses is, as a rule, greater than that of dry weight analyses.

## 2. Seasonal Variations

Seasonal variations of the energy values were not taken into consideration in Table 1. They can be important for material gathered at various times throughout the course of the vegetation period. This primarily concerns material from the *Trisetetum* and *Lolium multiflorum* plots and the litter from the forest stands. As an example, Table 2 shows the energy values of *Trisetetum* hay for various harvesting

Table 2. *Energy values (cal/g) of Trisetetum-hay at various harvesting dates*
(Values in parentheses = $s\%$)

| Date | unfertilized | | PK | | NPK | |
|---|---|---|---|---|---|---|
| | 1967 | 1968 | 1967 | 1968 | 1967 | 1968 |
| 27. 5. | | | | | | 4438 (1.0) |
| 3. 6. | | | | | 4474 (1.6) | |
| 18. 6. | | 4441 (0.7) | | 4372 (0.9) | | |
| 19. 6. | 4458 (0.6) | | 4396 (0.3) | | | |
| 2. 7. | | | | | | 4471 (0.3) |
| 11. 7. | | | | | 4400 (2.8) | |
| 3. 9. | | 4407 (1.1) | | 4326 (2.9) | | 4315 (3.8) |
| 5. 9. | 4380 (1.1) | | 4303 (2.2) | | 4405 (1.3) | |
| $\bar{x}$ | 4419 | 4424 | 4349 | 4349 | 4426 | 4406 |

dates throughout the years 1967 and 1968. The energy value of hay from the last date in both years was, as a rule, lower than that of the earlier ones. (The only exception was the last mowing of the NPK variant in 1967). This tendency is, however, not strongly pronounced. The differences between the various dates are so small that they are statistically significant in only a few cases. If the annual mean is calculated for every treatment, the energy values of the individual harvesting terms deviate only about 1% from these (1.3% at the most).

This finding is valid so far only for the example investigated, with several mowings during the vegetation period. Energy values given by Golley (1961) for an unmown, old field community, however, also show only slight seasonal changes.

The seasonal changes in energy values as found for the meadow plots are of the same magnitude as those found for the *Lolium multiflorum* plot. Leaves of *Fagus silvatica* showed somewhat greater energy value variations in 1968; but when sun and shade leaves were considered separately, the deviations from the annual mean did not exceed 2 %.

## 3. Influence of Different Fertilizer Treatment

The differences in energy value between the different fertilizer treatments of the meadow are also to be seen in Table 2. There is no apparent difference between hay from unfertilized and NPK-fertilized plots. Hay from the PK-fertilized plots had the lowest energy values in both years, but the difference between the annual mean values is very small (1.7% of the mean value of the PK treatment).

A comparison of these data with the dry matter yields demonstrates that the differences in energy fixation between the fertilizer treatments are attributable exclusively to differences in the dry matter yields. While, for example, in 1968 no differences in energy value were found between unfertilized and NPK plots, dry matter production, and hence energy fixation, of the unfertilized plots constituted only 30.5% of the corresponding values of the NPK plots.

# IV. Discussion

The differences in energy values between individual components of plant stands can be very great, but, when suitably fractionated, the differences within these components are much smaller. Therefore, energy values determined elsewhere can be taken over in many cases with only a slight risk of error.

The uniformity of the energy values as demonstrated in the present communication by, for example, the values found for different components of *Fagus silvatica* and *Picea abies* stands of various ages, shows that the results of studies on one stand are valid for all stands of the same type. In studies by OVINGTON and HEITKAMP (1960) on the energy values of different species of trees, among 11 coniferous species the greatest deviation from the mean was only 2.8% for needles and 1.6% for boles. 8 deciduous species showed corresponding deviations of 0.9% for leaves and 2.3% for boles. Results of research by GOLLEY (1961, 1969), however, demonstrated that the energy values of the various components of tropical trees are usually substantially lower.

This means that species which occur under comparable climatic conditions vary relatively little from one another in respect to the energy value of their important components, whereas comparison with species in other climatic zones yields greater disparities. Although the use of data from different species of trees presents a greater risk of error than if analyses are limited to a single species, this error is of relatively small scope as long as the disparities between deciduous and coniferous species, and between climatic regions are taken into account.

The energy values among grass species represented on the Solling Project plots vary only slightly. Between grasses and herbs, however, greater and quite significant differences occur. Hence, dry material from sites where grass is a primary component could differ substantially from dry material from sites where herbs dominate. Seasonal changes in the energy values of grasses, however, appear to be so minimal that analyses of material gathered at different times are comparable. According to the results of the studies on trees, however, more significant differences are to be expected when material from sites with considerably differing growth conditions are compared.

For the purpose of simplifying subsequent studies such as the already planned "Minimum Programs", the possibilities for utilizing energy values derived from elsewhere should be thoroughly clarified. The hitherto available data confirm the existence of such possibilities; a sound classification of the available energy values according to plant community, species and component will help to reduce substantially the error necessarily involved in taking over data from other sources. It is expected that the present comprehensive studies in the IBP will furnish the necessary foundations.

# References

Bliss, L. C.: Caloric and lipid content in alpine tundra plants. Ecology 43, 753 (1962).
Golley, F. B.: Energy values of ecological materials. Ecology 42, 581 (1961).
— Caloric value of wet tropical forest vegetation. Ecology 50, 517 (1969).
Lieth, H.: The measurement of calorific values of biological material and the determination of ecological efficiency. In: Eckard, F. E. (Ed.). Functioning of terrestrial ecosystems at the primary production level. Proc. Copenhagen Symp. 1968.
Ovington, J. D., Heitkamp, D.: The accumulation of energy in forest plantations in Britain. J. Ecol. 48, 639 (1960).
Roth, W. A., Becker, F.: Kalorimetrische Methoden zur Bestimmung chemischer Reaktionswärmen. Braunschweig: Vieweg u. Sohn 1956.

# M. Food and Energy Turnover of Leaf-eating Insects and their Influence on Primary Production

W. FUNKE

## I. Introduction

The significance of herbivores in a community is derived from the fact that:
a) they often use a considerable part of the organic substance produced by plants;
b) they diminish the photosynthetic production of plants;
c) being links in food chains, they serve other consumer groups.

With this manifold significance in view, my collaborators R. GRIMM, J. SCHAUER-MANN and I, as well as F. SCHWERDTFEGER and his co-worker K. WINTER, are investigating the commonest of the dominant leaf-eating insect species in two test areas in the beech forest* (B 1 a — beech stand about 120 years old, B 4 — beech stand about 60 years old); our investigations concern: 1. the herbivores' food and energy turnover and 2. their influence on the primary production of the beech tree.

## II. Food and Energy Turnover*

For the estimation of food and energy turnover of populations, the following data are collected (according to MACFADYEN, 1967; KLEKOWSKI, PRUS, and ZYROMSKA-RUDZKA, 1967; PETRUSEWICZ, 1967 e.g.):
a) Number and duration of stages throughout the year of life cycle
b) Instantaneous biomass of body (standing crop)
c) Cumulative total production of body, exuviae, eggs etc.
d) Cumulative cost of maintenance (respiratory losses)
e) Cumulative assimilation.

Similar to published investigations, e.g. of KLEKOWSKI et al. (1967), an energy budget for one individual (average of male and female) at all stages may be established as far as possible. Tabulating the cumulative elements of the energy budget in calories per unit of time, the energy flow may be obtained for each stage of development and, further, the efficiencies of production/assimilation, production/consumption and assimilation/consumption as well as the ratios of respiration/production etc. may be computed.

For an evaluation of productivity within the ecological system the data found must be related to actual populations in nature, i.e. to numbers of animals or, more exactly, to the abundance of the various stages affecting the natural abiotic conditions.

If a detailed life table of a population is constructed and a cumulative energy budget of an average individual of the species under consideration is established, a total budget may be computed for this population.

---

* Methods used to study energy flow see FUNKE and WEIDEMANN (chapter O).

# III. Species Used for Productivity Studies

Besides some Lepidoptera (chapter N) and the aphid *Phyllaphis fagi* L., it is the weevils with a number of species which are of outstanding importance in the sample areas.

They are *Rhynchaenus fagi* L.
*Phyllobius argentatus* L.
*Polydrosus undatus* F.
*Strophosomus melanogrammus* Forst.
*Otiorrhynchus singularis* L.

The imagines of all species eat leaves, *S. melanogrammus* also eats fallen foliage, and *O. singularis* various parts of plants. The larvae of *R. fagi* mine in leaves, the larvae of the other species live in the soil, feeding primarily on roots.

# IV. Methods of Measuring Population Density

Since phytophagous insects of forest trees are found in the canopy area, on the trunks, and on or in the soil — depending on their stage of development, the season and weather conditions — a number of different test methods are required (Balogh, 1958; Macfadyen, 1963; Klomp, 1966; Southwood, 1966 e.g.).

Some of them can at the same time be used for arthropods of other trophic levels and thus contribute further to the estimation of secondary production in general. Besides this they will give to a certain extent additional information on other fields of population and community studies.

## 1. Evaluation of Number of Viable Eggs Laid

The average number of eggs per female and the duration of oviposition period are determined in nearly all cases in the laboratory. In regard to natural conditions, the number of eggs per unit area and unit time must be calculated from the numbers of mature females.

## 2. Sampling from a Unit of Soil

The soil-living larvae and pupae are detected at present by square samples and direct sorting, partly with the use of Berlese-Tullgren funnels. In addition, an automatic extraction method developed by Kempson et al. (1963; see Weidemann, chapter P) and possibly some other methods (Southwood, 1966; Jackson and Raw, 1966) are to be used in the future.

## 3. Sampling Animals during Emergence

After emergence or hibernation on the ground, the imagines of most insect species migrate to the canopy area of the trees. On their way they are intercepted by various automatic catching devices, so-called *eclectors* (emergence traps, cf. Southwood and Siddorn, 1965; Southwood, 1966), which cover a more or less limited area. These devices take advantage of the phytophagous insects' orientation by light (positive phototaxis), gravity (negative geotaxis) and contours (tree trunks).

### a) Photo-Eclector of the Usual Shape and Function

The type most frequently used (Fig. 1a and 2) consists of a square wooden frame (side 1 m, height 0.5 m) with a pyramid-shaped roof of black cloth, supported by a frame of thin metal bars and fastened at the apex between a metal ring and a short

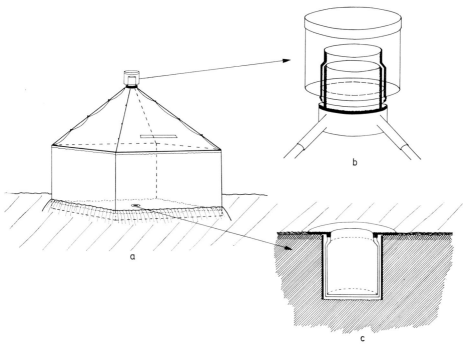

Fig. 1a—c.  Ground photo-eclector (emergence trap). a) total view; b) sampling box (light trap) with pipes and upper metal construction; c) pitfall trap (sectional diagram); particulars see text

plastic pipe ($\varnothing$ * 7.1 cm). Exactly fitted on this pipe — but easy to remove — there is a box of transparent polystyrene ($\varnothing$ 12 cm, height 9.5 cm) which serves as a light trap. The connexion between the box and the plastic pipe at the apex of the pyramid is another piece of pipe that has been glued into a circular hole in the bottom of the box (Fig. 1b). The inside of the wooden frame and the pipes is covered with black paint into which sawdust has been mixed in order to give climbing animals better opportunities of clinging. All positively phototactic (and negatively geotactic) animals fly or creep into the bright central collecting box, where they are killed and fixed in a mixture of picric acid and nitric acid (2000 cm³ of saturated picric acid + 15 cm³ concentrated nitric acid).

---

\* $\varnothing$ means diameter.

Two other types of eclectors with a frame of asbestos cement and plastic roofs have additional collecting boxes on all four sides, and at two or three different levels (Fig. 4).

All eclectors are dug into the mineral soil to a depth of several cm; moreover, earth several cm high is piled against the outside. Thus it is almost impossible for animals inside to get out or vice versa, so catch results can be attributed exactly to the enclosed area (usually 1 m²).

Fig. 2. Ground photo-eclectors in the beech stand

At the time when the forest floor is strongly exposed to sunlight, that is before the production of fresh leaves in spring, the catch boxes are screened from insolation by thin plastic disks, which are grey on top and white underneath and measure $20 \times 20$ cm. This prevents any sharp rise in temperature and concentration of the fixing liquid and possible maceration of the animals caught.

As the catch results are to reflect the total number of phytophagous imagines that emerge out of the soil area, early losses by predatory arthropods must be excluded as far as possible. This means that these animals must be intercepted on the ground as soon as possible after the eclector has been installed, or after their emergence. For this purpose every eclector is furnished with a pitfall trap (Fig. 1c). The walls of this trap consist of a plastic pipe (⌀ 7.1 cm; height 8 cm) dug into the ground; this prevents particles of the substrate from falling in and accommodates the catch jar (⌀ 5.6 cm, height 7.4 cm). The lid of the jar (twist-off cap) has a circular opening almost as wide as the opening of the jar. A ring of thin plastic (⌀ 12 cm) is fixed on the narrow rim of the lid; it rests on the pipe and is exactly on a level with the surrounding substrate; its upper surface is painted black and roughened by means of sawdust. The catch jar contains the same mixture of picric and nitric acid as the light trap. A zipper in the eclector cloth allows a quick exchange of pitfall trap jars.

In the placing of the eclectors the characteristic structural elements of the ground are considered (level areas, dips, elevations, tree stumps, places with and without herb layer). At several places 4 eclectors are connected around beech trunks, areas of special interest because of their intense soaking by the great quantities of water running down the trunks.

The microclimate (air temperature and relative air humidity) inside the eclector does not differ to any considerable extent from the climate in the stand during the summer months. Greater differences in respect to ground temperature must be expected on days of strong insolation before the production of fresh leaves. The ground covered by an eclector is less soaked by precipitation; there are no strong air movements, and the litter layer dries up more slowly. The most important differences, however, exist in respect of the lighting of the ground; light can get into the eclector through the central collecting box only. For lack of light, herbs cannot develop or prematurely die away. Eclectors with white cloths diminish this effect. Whether they give catch results as good as those with black cloths is being studied by my collaborator, G. STREY, in a beech forest that — unlike the Solling forest — has a rich herb layer. In the test areas of the Solling black cloths and wooden frames painted grey on the outside were preferred because visually striking marks could be a special attraction to other animals, and thus impair investigations concerned with these other animals. A certain attractiveness, however, could not be avoided; it was particularly strong only in the case of inhabitants of the lower trunk area (e. g. in the empidide *Tachypeza nubila* Meig.).

In order to exclude possible influences of the special conditions inside the eclector on the development of the insects in the soil, some eclectors are moved every week or every four weeks respectively. This procedure is significant in still another respect: by this means those insects can be included which are found in two or more generations per year.

Most eclectors are moved only once a year, at the time of snow break. Some of them remain at the same place, however, for at least 2 to 4 years, so we can find out which species need several years to develop in the ground or lie over as pupae. At the same time light is thrown on the general efficiency of the eclector, since some species of arthropods are quantitatively intercepted in the first year while others occur year after year; the latter must be able to propagate in the eclector (esp. a few species of Diptera) thus probably never being extracted completely. Further, the larval stages of Curculionidae and Elateridae seem to be able to immigrate annually beneath the eclectors' frame.

Most eclectors are kept ready for catching in winter as well. Simple roofs of firm cardboard or rush mats protect the cloths from damage, especially by heavy snow falls. From other eclectors the cloths are removed late in autumn; thus the ground inside the wooden frame is made accessible to winter weather.

The collecting boxes are checked every week with regard to the most important test objects and emptied every fortnight (in winter at longer intervals). Thus the catch results, besides yielding data for abundance calculations and information about the sex ratio, give information about the time of emergence at the same time, if we consider all species, laying the foundation for the characterization of the aspect sequence (Aspektfolge) of the community.

The efficiency of the eclectors was tested with various insects, particularly the most important test objects, the Curculionidae, but also some dominant species of Elateridae and Diptera. In every case between 90 and 100% of the individuals introduced were recaptured.

### b) The Extraordinary Significance of Photo-Eclectors

According to the results hitherto found, the eclectors are appropriate for an almost quantitative extraction of arthropods of all trophic levels in the trunk and canopy areas, as long as they have stages of development that live on the ground. Moreover they extract the imagines of detritophagous and carnivorous dipterous larvae, numerous inhabitants of the ground (esp. Carabidae and Staphylinidae) and, less quantitatively but nevertheless in large numbers, spiders and especially springtails. The sum total of all arthropods caught per m² in 1968 ranges between 3000 and 5000 individuals.

By measuring dry weights and measuring (or estimating) energy contents (Chapter O) of freshly emerged animals in respect of emergence abundances, we can carry out calculations, or at least approximately realistic estimations, of the *production of imagines (production of growth = Pg per ha per year [Pg-im./ha/yr])* of all common insect species that live above ground at least some of the time.

In view of the difficulties connected with the determination of the total animal production in a forest, the eclector method thus yields the most extensive data for the calculation of animal production at different trophic levels. The production of insect imagines (Pg-im./ha/yr), measured over several years, could be more useful in comparing different but similarly structured communities (e.g. different forests) than their total animal production, because the latter cannot be determined completely. For such comparisons, a standardization of the photo-eclectors (or emergence traps) would be desirable.

### c) Photo-Eclectors for Large Areas

The horizontal dispersion of most arthropods in the forest soil is unequal. This is the reason why, in addition to the usual "ground photo-eclectors" (see also Figs. 1 and 2), other novel collecting devices for larger areas were tested. The starting point was the fact that many animals, after emergence, use the tree trunks to migrate to the canopy. They can be intercepted on their way by means of so-called "arboreal photo-eclectors". These are funnels of black cloth with the large opening facing the ground; 3 or 4 of them are connected, thus forming a complete sleeve around the trunk about 2 to 3 m above the ground. The lower inner rim of every funnel is very close to the bark following its curve, while the outer rim — supported by a strong wire — protrudes. At the upper end of the funnel there is — as with the ground photo-eclector — a transparent collecting box containing fixing mixture (Fig. 3a and b). Two or three series of funnels one above the other improve the collecting result and make it possible in the case of some insects which cannot fly, to catch almost all the individuals climbing up the trunk (Fig. 4).

In stands with approximately equal distances from tree to tree, the catchment area of an eclector tree depends on the number of trees per ha; otherwise it must be calculated from the distances to the neighbouring trees.

Fig. 3a and b                                                  Fig. 4

Fig. 3a and b. Arboreal photo-eclector — four funnels grouped around a trunk a) side view; b) view from below (transverse section of the stem; small circles — light traps at the top of the funnels); particulars see text

Fig. 4. Arboreal photo-eclector in the beech stand; two series of funnels on top of each other (on the ground — ground photo-eclectors of an older type than that shown in Figs. 1 and 2; see IV 3a

Catch results of the arboreal photo-eclectors in the test areas in the Solling forest were extremely good but often — possibly under the influence of sun, wind and rain — showed striking differences in the different compass directions.

In the case of most species the catch results are not suitable for use in abundance calculations; there might be an exception with species which emerge within a short, limited period of time. The fact that the results are not completely satisfactory in this respect is due to numbers of animals being blown down to the ground by wind and rain everywhere in the stand; some of these will later be caught in the collecting boxes of eclector trees.

### d) The Significance of Arboreal Photo-Eclectors in Population and Community Studies

Even if arboreal photo-eclectors do not permit conclusions with regard to emergence abundances, they nevertheless yield some valuable information which could hardly be obtained otherwise, particularly in old beech stands. If the collecting boxes are emptied regularly (every week or every fortnight), very exact phenological data can be gathered about all inhabitants of the canopy and trunk areas and also about many species which are found only temporarily above ground, often in certain weather conditions only. Some information is of a similar kind — but more precise — as that yielded by ground photo-eclectors, e.g. first occurrence in the stand, or emergence time; furthermore, stage of development in the case of insect larvae and spiders, speed and duration of development, maturity, duration of activity period, life span in species with known bionomics.

In some cases the catch results provide clues as to the mode of changes in the numbers of individuals during the vegetation period and thus help in the interpretation of the abundance calculated by other methods (see below) at different times of the year.

Information is gathered about the "movements" of many species between the canopy and the ground. Thus on a single trunk (of 250 trees per ha) 1130 individuals of *Strophosomus melanogrammus*, a weevil which is unable to fly, were caught between April 23rd and August 10th, 1968 (that is, about 2% of the emergence abundance per ha). Some other beetles, Lepidoptera (both adults and larvae), Rhynchota, Diptera and spiders reached sum totals of similar size.

When several series of funnels on top of each other are installed as late as in summer, larger catches in the upper eclectors of animals which are unable to fly demonstrate that they run down the trunk (e.g. *Otiorrhynchus singularis*).

Furthermore, the significance of arboreal photo-eclectors lies in catching animals of different microhabitats, immigrants, animals which are passing through or passively brought in from adjacent biotopes.

Because of their comprehensive catch results, arboreal photo-eclectors yield a great many data on the aspect sequence in the stand during the year.

Besides numerous good qualities arboreal photo-eclectors, of course, have certain weaknesses, which must be considered in the evaluation of the catches. They might, in particular, form microhabitats and concentration centres for some species. They simulate refuges and thus, in some cases, catch strikingly large numbers of individuals in bad weather conditions and esp. in late autumn (e.g. *Rhynchaenus fagi*). On the other hand some species avoid the dark funnel region and often fly off the trunk when they reach it (e.g. *Phyllobius argentatus*). Funnels of transparent or white material would probably diminish these effects. So far they have not been used for reasons already mentioned (avoidance of visually striking marks).

Arboreal photo-eclectors are suitable for different types of forest. Catch results in a spruce stand were surprisingly good (e. g. at one tree about 40,000 individuals, all of a single aphid species were caught within two weeks). On the other hand, catches were extremely small in a beech forest with a dense herb layer. The causes of this are being investigated. Of special interest is whether this result is conditioned by the particular location, and to what extent inhabitants of the canopy which are unable to fly can exist at all in a forest with heavy space resistance („Raumwiderstand", HEYDEMANN, 1953) on the ground, and furthermore whether, in such a case, insects which can fly will take off from elevated starting points, e. g. herbs, and fly directly to the canopy.

## 4. Sampling in the Canopy Area in Respect of Fluctuations between Canopy and Ground Levels

### a) Census of Animals on Foliage Samples

With the mining stages of the weevil *Rhynchaenus fagi*, the following procedure is used. Between April and June, that is at the time it is developing inside the beech leaf, at intervals of 3—7 days, fixed numbers of leaves from different levels are collected from different trees; then eggs, larvae (different stages and sizes), pupae and adults ready for emergence are counted. The abundance of the different stages can then be calculated from the mean number of leaves per m² ground area (REICHLE and CROSSLEY, 1967).

This method is equally well suited to other leaf miners, to lepidopterous larvae feeding in leaf capsules (WINTER, chapter N) and to the determination of galls. But it is only with *Rhynchaenus fagi* that it yields data for the determination of the total consumption, production (Pg only), and the cost of maintenance from the egg to the imago ready for emergence, as I have emphasized in chapter O.

A different procedure is necessary with animals that live freely on leaves and twigs of the canopy area.

### b) Census of Animals in Shaking Samples

Many arthropods fall to the ground when there is strong shaking or vibration of the tree. This fact can be used for the determination of their numbers (SOUTHWOOD, 1966). Plastic sheets ($2 \times 10$ or $2 \times 5$ m) are spread on the forest floor. All trees whose branches extend over the covered area are then thoroughly shaken or banged with heavy mallets. The shake-down of animals and parts of plants is then carefully swept together and collected in plastic boxes. Every test area is treated in the same way several times in succession, until there are hardly any more animals coming down, except for the occasional aphids, cicadas, or other arthropods which cling well or are held by webs.

The samples are taken in at least two different places of a young beech stand (nearby B 4) at intervals of 1 to 3 weeks from April to October each time from a total area of 20 to 40 m².

The material collected is sorted immediately after sampling in the laboratory. The most important species are counted and weighed for the determination of abundance, stage and biomass. They are then used for assimilation and respiration measurements, for calorimetry etc. (FUNKE and WEIDEMANN, chapter O). By this simple method it is impossible to obtain the total number of animals. Therefore the whole sample, consisting of a large mass of plant material and the rest of the animals after the first sampling, is put into devices combining the properties of Berlese-Tullgren funnel and photo-eclector. In these a polystyrene box (serving as light trap) is mounted on the top of a conical lampshade. The lid of the box is pierced by a cable with a red bulb attached approximately in the middle of the shade. The light is switched on and off at 6-hour intervals in order to remove the animals from the wire gauze sieves with the least possible damage. The collecting boxes are illuminated from above. Using this method, it is possible to collect at the same time animals avoiding warmth, dryness and light and positively phototactic species.

Catch results depend very much on weather conditions. Large numbers of individuals were caught during prolonged dry spells with light winds and moderate temperatures. Local weather conditions change so fast that even catches made at short intervals differ so much that no conclusions can be drawn about population density when only this method is used.

At any time some of the inhabitants of the canopy are on the ground. By installing ground photo-eclectors immediately after every sampling the number of these animals can be determined — at least on a few m² — and be included in the abundance calculation later on. It must be observed that some weevils, in particular, do not appear in the collecting box until several weeks later and at certain times perhaps cannot be caught quantitatively at all. A practically indeterminable number of animals remains in the canopy in spite of the knocking of the trunks. This means that an accurate determination of the abundance is impossible — apart from the difficult problem of dispersion and its dynamics. The catches of arboreal eclectors cannot be of any great help, either. They only give clues, as mentioned above (IV 3d), as to the mode of change in the number of individuals of a population.

Thus shake-sampling and moved ground photo-eclectors can give only minimum abundances. It can even happen that a short time after the emergence of a beetle population lower data are calculated than at a later time. It is obvious that in such a case in calculating biomass, the amount of assimilation and the losses by respiration of the population, the larger number must be used for the earlier point of time as well. Even using the higher number does not do justice to the actual state of affairs at the earlier point of time; the number at that time should be larger, but we do not know by how much. The fact that energy turnover is calculated from data that are necessarily too low is compensated to a certain extent — regarding the weevils mentioned above (III) — in the following way. The amounts of food and respiration are normally (except for fast developing larvae) related to the number of individuals and biomass data per month. At the time of emergence, emergence abundance and emergence biomass are used for calculations of a month or half a month. The results thus obtained are too large because the decrease in the number of individuals, which set in at once, cannot be taken into consideration.

A special difficulty arises in the case of *Rhynchaenus fagi*. Beginning in summer many beetles migrate into adjacent biotopes*. Starting in August 1968 more and more

---

* See e.g. SCHINDLER (1965).

of them were "knocked" in a spruce stand, which is completely surrounded by large
beech stands of different ages. From the end of September onwards — according to
the tests of my collaborators — they were even more frequent in the spruce stand than
they had ever been in the young beech stand (area B 4). Knocking samples in the
spruce stand at regular intervals provide information about changes in the number
of individuals only for the part of the population hibernating there. More detailed
statements about the density of the total population cannot be made until spring,
when the beetles return to the test areas. For the determination of the energy losses
by respiration of the population in the preceding months, we must again rely on the
small spring abundances.

### c) Calculation of Population Density in Old Beech Stands

Height and diameter of the trees in old beech stands do not permit the shake-
sampling done in young beech stands and spruce stands. For this reason changes in
the numbers of individuals during the course of the year can only be roughly estimated.

In heavy rains and violent winds numerous animals drop to the ground. Ground
photo-eclectors can be moved to a new place; the individuals caught in the light
traps give at least some clue as to the magnitude of different populations. A further
clue can be found by comparing the knocking samples in the young beech stand and
the catches of arboreal eclectors in both test areas; in doing so, however, differences
of the density of the stand, i.e. of the catchment area of each trunk, must be considered
(see IV 3c). In order to get a first estimate we can — ignoring the fact that the mortality factors do not have exactly equal effects—assume that the numbers of individuals
after their emergence change in approximately the same proportion in the old stand
(B 1a) as they do in the young stand (B 4).

# V. Preliminary Results in the Study of Production and Energy Turnover

## 1. Production of Imagines

Regarding the production of imagines (Pg-im./ha/yr) for the weevils mentioned,
we obtain for both sample areas (B 1a and B 4) the following values based on the
emergence abundance gained by the ground photo-eclectors in kg (dry weight)/ha
for two years (1968, 1969):

| B 1a/1968 | B 1a/1969 | B 4/1968 | B 4/1969 | |
|---|---|---|---|---|
| 2.292 | 1.910 | 0.714 | 1.453 | kg/ha/yr |

For the representation in kcal/ha/yr a provisional value of 5.5 kcal/g dry weight
was assumed for all species (see KACZMAREK, 1967; GÓRECKI, 1968).

| B 1a/1968 | B 1a/1969 | B 4/1968 | B 4/1969 | |
|---|---|---|---|---|
| $12.6 \times 10^3$ | $10.5 \times 10^3$ | $3.9 \times 10^3$ | $8.0 \times 10^3$ | kcal/ha/yr |

In comparing the results, marked differences are found between the two areas
under study.

According to the eclector catches, the share of Curculionidae imagines in the whole production of phytophagous insects (imagines only) was about 80% for B 1 a in 1968. Their share of the total production of insect imagines was roughly estimated to be 30—40%.

## 2. Cost of Maintenance in Imagines

With the methods described for calculating the abundance and mean biomass per month and taking into account the seasonal temperature average (KIESE, chapter S), the following energy losses by respiration (assuming an oxycaloric coefficient of 4.8 kcal/1 $O_2$) were obtained from the investigations of my collaborators for the weevils mentioned (with the exception of *Otiorrhynchus singularis*):

| B 1a/1968 | B 1a/1969 | B 4/1968 | B 4/1969 | Respiration of imagines |
|---|---|---|---|---|
| $16.5 \times 10^3$ | $14.6 \times 10^3$ | $4.0 \times 10^3$ | $6.0 \times 10^3$ | kcal/ha/yr |

The annual loss of energy in the community is thus at least of the same order of magnitude as the energy incorporated in the just-emerging imagines of all 5 species together.

The values obtained for respiration in B 4 are too low; this is due to the high abundance of *Otiorrhynchus singularis* which was not included in the respiration measurements.

As regards energy flow, the Curculionidae populations (beetles only — production of imagines and respiration of imagines) reached an amount of at least 0.08% of the primary production of the old beech stand (wood and leaves — B 1/1968). Even if the true value were twice as great, the uncertainty is thus not large related to the ecosystems energy budget.

# VI. The Influence of Leaf-eating Insects on Primary Production

Phytophagous insects diminish the photosynthetic production of plants. Often the extent of the loss so caused cannot be judged by the amount eaten so much as by the place of the attack, the stage of development of the plant or its organs, and the season of the attack. Thus, eating the petiole is more damaging than making holes in the lamina. Eating, sucking or the deposition of eggs frequently cause deformations (crippled growth) of young leaves (BALOGH, 1958; multiple and far-reaching effects see VARLEY, 1967). The destruction of leaves in spring will show an effect throughout the vegetation period, while leaf eating in autumn is comparatively harmless.

Of the manifold influences of phytophagous insects on the production of the beech, only the loss caused by leaf eating can be determined at present.

Since all tests are to be related to the production phenology of the beech, foliage samples must be taken at regular intervals (from different levels in the canopy and from different places in the stand).

Leaves collected are counted. Their surface eaten away is estimated or measured in random samples (FUNKE and WEIDEMANN, chapter O) taking into account the thickness of the leaves. Food quantities (g dry weight/m²) can then be calculated from the mean number of leaves (sun and shade leaves), their surface and dry weight per m² ground area (HELLER, chapter B).

By means of photosynthesis data in relation to the dry weight of leaves (SCHULZE, 1970; LANGE and SCHULZE, chapter A) the diminution of primary production by phytophagous insects might then be estimated up to a certain degree.

# References

BALOGH, J.: Lebensgemeinschaften der Landtiere. Berlin: Akademie-Verlag 1958.

GÓRECKI, A.: Calorimetry in ecological studies — In: Methods of ecological bioenergetics — Handbook of IBP Training Course in Bioenergetics pp. 125—133. (Ed. W. GRODZIŃSKY, and R. Z. KLEKOWSKI) Warszawa-Krakow: Polish Acad. Sci. 1968.

HEYDEMANN, B.: Die Biotopstruktur als Raumwiderstand und Raumfülle für die Tierwelt. Verh. Deutsch. Zoolog. Ges. Hamburg 1956 Zool. Anz. (Suppl.) **20**, 332—347 (1957).

JACKSON, R. M., RAW, F.: Life in the soil. Studies in Biology, no. 2. London: E. Arnold 1966.

KACZMAREK, W.: Elements of organization in the energy flow of forest ecosystems (preliminary notes). (Secondary productivity of terrestrial ecosystems, pp. 663—678. Ed. K. PETRUSEWICZ) Warszawa-Krakow: Polish Acad. Sci. 1967.

KEMPSON, D., LLOYD, M., GHELARDI, R.: A new extractor for woodland litter, Pedobiologia **3**, 1—21 (1963).

KLEKOWSKI, R. Z., PRUS, T., ZYROMSKA-RUDZKA, H.: Elements of energy budget of *Tribolium castaneum* (Hbst) in its developmental cycle. (Secondary productivity of terrestrial ecosystems, pp. 859—879. Ed. K. PETRUSEWICZ) Warszawa-Krakow: Polish Acad. Sci. 1967.

KLOMP, H.: The dynamics of a field population of the pine looper *Bupalus piniarius* L. (Lep. Geom.): Advanc. Ecol. Res. **3**, 207—305 (1966).

MACFADYEN, A.: Animal ecology: Aims and methods, 2. ed. London: Pitman 1963.

— Methods of investigation of production of invertebrates in terrestrial ecosystems. (Secondary productivity of terrestrial ecosystems, pp. 383—412. Ed. K. PETRUSEWICZ). Warszawa-Krakow: Polish Acad. Sci. 1967.

PETRUSEWICZ, K.: Concepts in studies on the secondary productivity of terrestrial ecosystems. (Secondary productivity of terrestrial ecosystems, pp. 17—49. Ed. K. PETRUSEWICZ). Warszawa-Krakow: Polish Acad. Sci. 1967.

REICHLE, D. E., CROSSLEY Jr. D. A.: Investigation on heterotrophic productivity in forest inset communities, pp. 563—587. Warszawa-Krakow: Polish Acad. Sci. 1967.

SCHINDLER, U.: Zum Massenauftreten des Buchenspringrüßlers 1963 und 1964. Forst- u. Holzwirt **20**, 1—5 (1965).

SCHULZE, E. D.: Der CO₂-Gaswechsel der Buche (*Fagus silvatica* L.) in Abhängigkeit von den Klimafaktoren im Freiland. Flora **159**, 177—232 (1970).

SOUTHWOOD, T. R. E.: Ecological methods. London: Methuen 1966.

— SIDDORN, J. W.: The temperature beneath insect emergence traps of various types. J. Anim. Ecol. **34**, 581—585 (1965).

VARLEY, G. C.: The effects of grazing by animals on plant productivity. (Secondary productivity of terrestrial ecosystems, pp. 773—778. Ed. K. PETRUSEWICZ) Warszawa-Krakow: Polish Acad. Sci. 1967.

# N. Studies in the Productivity of Lepidoptera Populations

K. WINTER

## I. Introduction

Among the primary consumers in forests Lepidoptera, i.e. their larvae, are of considerable significance in the diminuation of the primary production of trees and consequently in the energy flow of the trophic level they belong to.

The investigations aim at tracing the path of the energy which is drawn from the primary products of the beech by caterpillars. For this purpose abundances, the production of body substance including exuviae and secretions, energy losses by respiration, and the amount of assimilated food must be determined and be investigated as to their energy contents.

The investigations are supervised by F. SCHWERDTFEGER and carried out in collaboration with W. FUNKE and his co-workers in two beech stands of different age (B 1a: 120 years old, B 4: 60 years old) in the Solling IBP areas.

## II. Methods

The methods used are basically the ones described by FUNKE (chapter M) and FUNKE and WEIDEMANN (chapter O). Therefore only a short survey and some additions specific for lepidoptera will be given here. The basic requirements in studies of productivity are investigations about population dynamics and the food and energy turnover of average individuals with regard to all stages of development and maturation.

### 1. Population Dynamics

Research in population dynamics in respect of production ecology is more exactly a study of abundance dynamics (SCHWERDTFEGER, 1968), e.g. the fluctuations in population density by natality and mortality, immigration and emigration.

Therefore abundances (ind./ha) must be checked continually. Since different stages of a population are found at different habitats in the forest various methods must be used. Data obtained are listed in life tables.

#### a) Abundance of Eggs

The abundance of eggs was calculated by the average number of eggs deposited per female in the laboratory and the density of females per unit area.

#### b) Abundance of Caterpillars

##### $\alpha$) Shaking Samples

Most species inhabiting the canopy area of young beech trees are caused to fall to the ground by strongly shaking or knocking the trunks. In this way they can be projected on a plastic sheet of 10 or 20 m² spread out on the ground under the test trees (see Funke, chapter M IV 4b).

The number of caterpillars "knocked" varied in the tests. Apart from the actual density changes, climatic conditions such as precipitation, wind, and temperature exert an influence on catch results. Whenever in a test, after hatching of all eggs, fewer individuals were caught than in a later test, abundance calculations were based on the higher figure, because a population can only decrease in the course of the year. In such cases the earlier lower figures must be due to temporary climatic conditions. For considerations and investigations concerning the estimates of abundances of animals inhabiting the canopy area see Funke (chapter M).

### β) Foliage Samples

Caterpillars of some species live in leaf mines (e. g. *Lithocolletis faginella* Z., *Nepticula basalella* Hs.) or leaf webs (e.g. *Chimabacche fagella* Schiff., *Acleris* spec., *Pandemis corylana* F.). Only a small number of them fall down in knocking tests, so that a different method has to be applied. Branches were taken from different trees at five different levels, 5000 leaves were counted and the caterpillars attached to them collected. Their abundance can be determined by means of the leaf area index (Heller, chapter B).

In the old beech stand (B 1a), the height and the diameter of the trees do not permit shaking samples, so that only random samples could be taken occasionally by means of special climbing irons (Swiss "Baumvelo").

### c) Abundance of Pupae

Pupae of Lepidoptera generally live in the litter or humus layer, and are often fixed between dead leaves or other materials; so they are hardly detectable. Because of the high mortality of the eggs and caterpillars the abundance of pupae is relatively low; moreover the dispersion varies considerably. For these reasons it would be necessary to take numerous square samples and search them for pupae by hand. In practice this method is not only too laborious and harmful in respect of the total community but moreover would yield very doubtful results. Therefore abundances of pupae are calculated from the last census of larvae and the density of emerged adults. The mortality during this time, represented graphically, is assumed to be linear.

### d) Abundance of Imagines

Abundance of imagines emerging on the soil surface is estimated from catch results of photo-eclectors. "Ground photo-eclectors" (see Funke, Fig. 1 and 2) are very effective with species which emerge in large numbers/m² and are dispersed rather homogeneously. "Arboreal photo-eclectors" are collecting devices for larger areas (Funke, Fig. 3 and 4). Their catch results are suitable for use in abundance calculations especially in the case of species which emerge in a short period of time and are unable to fly (e.g. females of *Chimabacche fagella* Schiff., *Erannis defoliaria* Cl. and *Operophthera fagata* Hb.). The number of trees per ha is the base by which the mean catchment area of photo-eclector trees is computed and the density of emerging Lepidoptera is estimated. Arboreal photo-eclectors seem to give better results in the beech stands than do grease bands, which were used too. Moreover these devices are of far-reaching interest in population studies (Funke, chapter M IV 3d), because many caterpillars, which have dropped everywhere to the ground from various causes are caught partially on their way up the photo-eclector tree throughout the season. Important information is gained about phenology as to the date of emergence and the duration of developmental stages.

## 2. Food and Energy Turnover

Food and energy turnover of populations are calculated from the energy income (assimilation of food, A) used for transformation to body tissue (production, P) and heat lost in vital processes (respiration, R) in a defined period of time.

Assimilation may be computed by two equations:

A = C — FU (where C means consumption and FU rejecta) and

A = P + R.

The values of assimilation in both cases must be equal:

C — FU = P + R.

In order to make a statement about the energy balance all parameters mentioned must be determined (defined in detail by Funke and Weidemann, chapter O, according to Macfadyen, 1967, and Petrusewicz, 1967).

The investigations must take into account to what degree results obtainable in the laboratory can be transferred to natural conditions without too large errors.

### a) Determination of Production

For the calculation of production there are two main prerequisites. First, dry weights and energy contents of all stages sampled in the stand must be determined (methods used are described by Weidemann, chapter P). Second, by using the abundances and the specific caloric values of the average individual stages the standing crop of the population (kcal/ha) is calculated for each time of sampling. Considering the mortality in the population and assuming immigration balanced by emigration, we may compute production by summing up the energy contents of

1. standing crop at a fixed moment;

2. individuals which died previously;

3. exuviae and secretions of all individuals, no matter whether they are alive or not;

4. reproductive products, especially eggs.

1—3 are considered to be parts of the production of individual growth (Pg), 4 to be the production of reproduction (Pr, Petrusewicz, 1967). In some cases, however, lepidoptera do not feed after emergence; the energy content of Pr then already must be contained in Pg totally and we should not sum it up once more.

### b) Determination of Respiration

A population's energy losses for cost of maintenance are evaluated by the gathering of data from all developmental stages and from males and females at different states of maturation. The investigations are carried out according to Funke and

WEIDEMANN (chapter O) with the manometric principle of Warburg apparatus in connexion with the cooling thermostat Frigomix. The different sizes of the stages made four types of reaction flasks necessary, ranging in capacity from 1 ml to 30 ml. All apparatuses and flasks were made by Braun (3508 Melsungen, Germany).

The test material consisted of eggs and pupae bred in the laboratory and larvae and imagines brought in from outdoors. Since the respiration rate changes in the course of development of the individual it had to be determined for all stages and both sexes. Furthermore it had to be taken into account that $O_2$-consumption varies with different states of development of the same stage.

The average temperature — at 2 m above the ground — of the respective month in the preceding year served as the test temperature.

### c) Determination of Assimilation

In laboratory cultures caterpillars were fed on fresh beech foliage, which was usually renewed every day. Then the remaining food was dried. Leaf samples of exactly the same quantity (fresh weight) and quality as those which were used for feeding were treated in the same way. By the differences of dry weights the consumed food quantities could be determined (VARLEY, 1967). The faecal pellets were collected every day, dried and weighed. Representative samples from each larval stage were stored up for later calorimetric measurements.

Food intake and rejecta are determined for each stage. Therefore the assimilation of one average individual at each state of development can be computed, and — taking into account population dynamics — the food assimilation of the total population can be calculated.

## III. Some Results

About 40 species of Lepidoptera are observed in the test stands, most of them are phytophagous on beech trees.

During two years of research six species are considered to be predominant primary consumers in the stands. They are *Chimabacche fagella* Schiff. (Gelechiidae), *Acleris sponsana* (Tortricidae), *Ennomos quercinaria* Hufn., *Erannis defoliaria* Cl. (Geometridae), *Operophthera fagata* Hb. (Geometridae) and *Colocasia coryli* L. (Noctuidae). Only three species of them are intensively investigated in the laboratory. The share in primary consumption of the other species is estimated as well as possible, using data of similar species with well known life cycles (e. g. *Operophthera brumata* instead of *O. fagata*; see VARLEY, 1967).

The dominance of each species and thus its significance within the community can be quite different in different years; the same holds true as to the ecological importance of the total group.

In 1968, e. g., *Ch. fagella* was dominant over all other species while in 1969 it was far less significant. Its standing crop biomass in August amounted to 36% (B 1a) and 19% (B 4) of the values of the preceding year. Nevertheless *Ch. fagella* still ranked second behind the roller moth *A. sponsana*. The populations of the two species made up about 80% of the total Lepidoptera production.

Up to this time the investigations regarding food and energy turnover of lepidopterous populations are performed mainly with *Ch. fagella.*

The energy equivalents of the oxygen consumed, of body tissue and excrements
and the dependence of these equivalents on factors of development or season could
not yet be determined. So the following estimates are based on values taken from
caterpillars of *Hyphantria cunea* Drury fed on *Acer negundo* (GERE, 1957). Thus the
following assumptions were made:

body tissue  = 5.4 kcal/g dry weight

food *(Acer)* = 4.3 kcal/g dry weight

faeces        = 4.2 kcal/g dry weight.

The energy content of beech leaves is about 5 kcal/g dry weight (RUNGE, chapter
L). The value of *Acer* was temporarily used because energy contents of food and faeces
are connected.

The oxycaloric coefficient of 4.8 kcal/l $O_2$ (SLOBODKIN, 1962) is applied by several
other investigators in secondary production. Dry weights of exuviae and secretions
(webs) could not be determined.

The calculations of energy flow through the population of *Ch. fagella* contain
several inaccuracies and must at this time be considered as a very rough approximation
to the actual facts.

Using the two possible ways of determining assimilation $(A = P + R$ and
$A = C - FU)$ we obtain the following data for the research areas B 1a and B 4
in 1969 (see Table 1):

Table 1. *Assimilation of Chimabacche fagella* (kcal/ha/yr)

|  | B 1a | B 4 |
|---|---|---|
| Production (P) | $4.3 \times 10^3$ | $1.6 \times 10^3$ |
| + Respiration (R) | $+ 10.6 \times 10^3$ | $+ 4.6 \times 10^3$ |
| = Assimilation (A) | $= 14.9 \times 10^3$ | $= 6.2 \times 10^3$ |
| Consumption (C) | $74.0 \times 10^3$ | $27.0 \times 10^3$ |
| — Rejecta (FU) | $- 48.3 \times 10^3$ | $- 18.0 \times 10^3$ |
| = Assimilation (A) | $= 25.7 \times 10^3$ | $= 9.0 \times 10^3$ |

The population in the old beech stand took in with its food about 0.5% (1968:
3.5%) of the amount of energy contained in the mass of leaves produced here per ha
and year.

In further investigations the discrepancies in the results of the two computations
will be cleared away. It is certain that there are striking differences in the results of
the two test areas, as it was with the research on weevils (FUNKE, chapter M V).

# References

GERE, G.: Untersuchungen über den Energieumsatz der Raupen von *Hyphantria cunea* Drury. Acta Zool. Acad. Sci. Hung. **3**, 89—105 (1957).

MACFADYEN, A.: Methods of investigation of production of invertebrates in terrestrial ecosystems. (Secondary productivity of terrestrial ecosystems, pp. 383—412. Ed. K. PETRUSEWICZ). Warszawa-Krakow: Polish Acad. Sci. 1967.

PETRUSEWICZ, K.: Concepts in studies on the secondary productivity of terrestrial ecosystems. (Secondary productivity of terrestrial ecosystems. Ed. K. PETRUSEWICZ). Warszawa-Krakow: Polish Acad. Sci. 1967.

SCHWERDTFEGER, F.: Ökologie der Tiere II, Demökologie. Hamburg-Berlin: Parey 1968.

SLOBODKIN, L. B.: Energy in animal ecology. Advanc. Ecol. Res. **1**, 69—101 (1962).

VARLEY, G. C.: Estimation of secondary production in species with annual life cycle. (Secondary productivity of terrestrial ecosystems, pp. 447—457. Ed. K. PETRUSEWICZ). Warszawa-Krakow: Polish Acad. Sci. 1967.

# O. Food and Energy Turnover of Phytophagous and Predatory Arthropods

## Methods Used to Study Energy Flow

### W. Funke and G. Weidemann

## I. Introduction

"The behavior of energy in ecosystems can be conveniently shorthanded as energy flow . . ." (Odum, 1968).

The total energy income (assimilation, A) of a community, a trophic level, a population, or a single individual, is either stored as potential energy (production, P) or used for maintenance (dissipated by respiration, R). Thus energy flow (E) can be written as

$$E = A = P + R$$

according to Lindeman (1942), Richman (1958), Petrusewicz (1967), Odum (1959, 1968).

The above-mentioned equation makes it clear which parameters must be measured before the energy budget of a population can be made up: production, respiration, and assimilation.

## II. Production

According to Petrusewicz (1967) and Macfadyen (1967) production is the amount of organic material, or the potential energy contained in it, including offspring, body growth and secreta, if produced (e. g. spiders' webs, cocoons), elaborated by the population of a defined area during a defined period of time, whether the whole population survives to the end of that period or not. Hence production may be expressed as

kg (live, dry, or ash-free weight)/ha/year, or as kcal/ha/year.

The latter expression is needed if energy flow is to be considered.

Since the methods for the evaluation of production are highly dependent on the biology and population dynamics of the species concerned, they are treated by Funke (chapter M) and Winter (chapter N) for phytophagous insects and by Weidemann (chapter P) for predatory arthropods of the ground layer.

Conversion of biomass data to energy equivalents requires measurement of the energy content of all developmental stages of a species, if necessary at different times during the year. Measurements are done with a microbomb calorimeter after Phillipson (1964) (now available from Gentry and Wiegert Instruments Inc., Aiken, South Carolina, USA) in combination with a potentiometer recorder (Honeywell Electronics 194 chart recorder).

# III. Respiration

Respiration is measured as the oxygen consumption (for exceptions, see section III 1f) of single individuals in a Warburg apparatus, type V 166 (or VL 166) of the firm B. Braun (Apparatebau, D-3508 Melsungen, Germany) in connection with the cooling thermostat Frigomix of the same firm. Different respiratory flasks are used according to the size of the animals to be studied.

Since the respiratory measurements will be used to evaluate the energy loss of a population under natural conditions, some requirements must be fulfilled to render that possible. The oxygen consumption of all developmental stages and both sexes has to be measured taking into account life cycle and duration of stages (PHILLIPSON, 1962). The abiotic conditions (temperature, humidity, light regime) must be similar to those in the natural habitat. Differences in the biology and physiology of phytophagous and zoophagous arthropods require modifications of the methods used.

## 1. Phytophagous Insects (FUNKE)

### a) Animals Used for Measurements

Respiration data are measured mainly with animals from the test area (methods of capture see FUNKE, chapter M). Subjects which are difficult to catch in the winter months, are put into cages in autumn, but kept under entirely natural conditions otherwise, and thus are easily accessible at any time. Soil-dwelling and root-eating larvae must be obtained from outdoor cultures as a rule. If eggs and young larvae cannot be found easily in the field, they must be taken from laboratory cultures, too.

### b) Respiratory Flasks

The respiratory flasks are usually conical. They have no sidearms; this is advantageous with regard to the small respiration volumes and, above all, in providing no hiding places for climbing animals and thus avoiding long diffusion distances. The small inner KOH-vessel is covered with fine gauze to protect the animals. Particularly accurate measurements are required for small animals, since their oxygen consumption must be derived from short-period experiments. Therefore micro manometers and micro reaction vessels (volume about 0.7 ml) are used in such cases (ZINKLER, 1966).

### c) Experimental Conditions

Many phytophagous insects, especially larvae, eat almost incessantly apart from pauses caused by moulting. Lack of food for a long time (in the reaction flask) would distort the normal respiration rate. For this reason, food is taken away from these very active eaters only shortly before measurement begins. Before starting the test, all animals are kept at the test temperatures for at least 12 hours, usually longer. In order to allow animals to acclimatize to the light and general conditions in the respiration flask, they are put into the flask at least one hour before measurements are started.

Locomotor activity, and hence the respiratory metabolism of herbivores and other arthropods in the canopy area, is probably influenced by light, especially when food is lacking. For this reason similar light conditions for all animals in a test series are created by means of a photo-Warburg apparatus (type VL 166). By regulating the light intensity, not only day and night but also dusk and dawn periods may be simulated. Thus a division into active and resting periods depending on the light may be possible.

Most measurements are done at temperatures corresponding roughly to the mean ground or air temperature at the appropriate season, depending on whether the various development stages are found in the soil or above ground. Close correspondence to outdoor temperatures (exact monthly or weekly averages in various strata) is aimed at in exceptional cases only (e. g. soil-living larvae).

Such a close correspondence is normally of little value, especially with most stages living above ground for the following reasons: a) During the vegetation period the phyllophagous insects in a forest, with the exception of leaf miners and gall inhabiting forms, are always found in the canopy area, the trunk area, and on the ground. The proportional distribution of the insects in these three strata changes and unfortunately cannot be determined exactly (see Funke, chapter M). On sunny days the animals differ in the intensity and duration of their exposure to the sun. Thus they are subject to quite different temperatures which, moreover, may change rapidly. b) The abundance of a population cannot be exactly estimated (see chapter M), thus a really exact calculation of the total respiration of a population per ha and year is impossible.

Although all strata are considered, the respiration data are measured at test temperatures that vary considerably at times from field temperatures; the calculation of the total respiration made from them will therefore be in error. The error, however, is small in comparison with that arising from our incomplete knowledge of abundance and population dynamics.

### d) Duration of Measurements

The daily oxygen consumption is the basis for the calculation of energy losses. As respiration is subjected to changes caused by diurnal physiological rhythms, the determination of daily consumption normally requires measuring periods of at least 24 hours (Phillipson, 1962). For the above-mentioned reasons this cannot be done in all cases. Therefore daily consumption must often be derived from short-period measurements in light and dark. In such cases it would be desirable to measure the respiration rate only when it is near the hourly average of a 24-hour measurement. Because this is not practicable, full data are given about time and duration of measurements, experimental conditions and all previous treatments.

### e) Repetition of Measurements

In measuring respiration data not only all stages of development and different sizes must be considered, but also changes in the physiological state and their dependence on the season (Phillipson, 1962, e. g.). Since these changes take place rapidly at some times and more slowly at others, measurements on the different subjects (eggs, larvae, pupae, adults) are carried out at intervals ranging from 12 hours (eggs, pupae) to four weeks.

## f) Determination of Respiratory Loss in Leaf Miners

A special new method for the determination of energy losses by respiration seems to be practicable in leaf-mining insects. In the process of energy transformation during a defined period of time, expressed by the equation (after PETRUSEWICZ, 1967)

$$C = P + R + FU,$$

where $C$ = consumption or food intake, i.e. the energy content of material ingested,
$P$ = production, i.e. the energy content of all organic matter produced, composed by $Pr$ (= production from reproduction, i.e. newly produced organisms) and $Pg$ (= production from body growth, including exuviae, secretions),
$R$ = respiration or cost of maintenance, i.e. the energy required for the support of life,
$FU$ = rejecta, including faeces and excretory products, i.e. the energy content of material egested

all materials remain in the mines. The only component lost is $R$. Therefore it must be possible to determine $R$ by measuring $C$, $P$, and $FU$.

Leaf parts with mines containing the animals, all their faeces, excreta, and all substances produced up to the time of sampling (i.e. $Pg + FU$), are dried and subjected to calorimetry.

Leaf parts of exactly the same size and thickness (if possible from the same leaves) but without mines, therefore including energy content of $C$, are treated in the same way. The loss of energy by respiration is calculated from the difference between the leaf samples under the following assumptions: a) the energy content of the leaf tissue surrounding the mine and in the shrivelled upper and lower epidermis of the mine remains constant; b) faeces have not suffered premature bacterial degradation.

The result, however is not entirely correct, for one component in particular, not included in PETRUSEWICZ's equation, must be taken into consideration, i.e. in the mined leaf parts the energy increment due to freshly deposited eggs. Although their energy content in relation to that of the whole sample is very low and can perhaps be only roughly estimated, we may regard it as follows:

| energy of respiration of the mine inhabitant | = energy content of leaf part without mine | — energy content of leaf part with mine | — energy content of freshly deposited egg |
|---|---|---|---|

or in terms of the above mentioned equation

$$R = C - (P + FU) - OV$$

where $P$ in this special case = $Pg$, i.e. the energy content of organic matter only produced by body growth and $OV$ = a single deposited egg and the energy contained in it.

This method should have several advantages over all other respiration measurements so far used: the animals remain under the natural, varying temperature and light conditions in the study area up to the time of testing or, more exactly, the time of drying. It should be possible to determine the total energy loss by respiration in the time of sampling at every developmental stage up to the emergence of the imago.

Making allowance for mortality up to the imaginal stage (Funke, chapter M), the loss of energy is thus calculated for the whole population in a more simple and exact way than from direct respiration measurements. The method may also be used for other endophagous animals, such as wood-boring or seed-eating insects.

## 2. Zoophagous Arthropods of the Ground Layer (Weidemann)

### a) Animals Used for the Measurements

Animals to be used for measurements of respiration are captured in the study area by hand or pitfall trapping, or are taken from laboratory cultures (eggs, larvae). As a rule they are fed in the laboratory to bring them all to the same degree of satiation. They are then put on a food-free substrate (damp cellulose) for defecation where the temperature is maintained near the intended experimental temperature.

### b) Experimental Temperature and Humidity

Oxygen consumption of poikilotherms is largely influenced by the temperature of the surrounding medium. The choice of temperature for the experiments is therefore very important, especially with the Warburg apparatus, which works at constant temperatures.

Unlike Phillipson (1962, 1963), who worked with litter-inhabiting harvest spiders at 16° C throughout his study period from May to November, we conduct our measurements at the monthly mean temperatures of the litter layer.

This ensures that oxygen consumption of, e.g. a larval stage, which lasts only three weeks in early spring, is measured at the appropriate temperature. One disadvantage of this procedure is, that comparisons cannot be made between stages in respect of respiratory rate, unless additional measurements are possible. But, as the purpose of our measurements is to compute the energy loss under natural conditions, this is not important.

Humidity is kept above 90% RH, a value necessary for most inhabitants of the forest litter layer, by means of a strip of damp filter paper. This prevents excessive energy loss due to transpiration, which may be very high, as was shown by Schmidt (1956) with Carabidae.

The animals are left for 6 to 12 hours in the respiratory flasks for adaptation, both to the experimental temperature and to the flask.

### c) Duration of Measurements

Oxygen consumption is influenced not only by temperature but also by activity. Motor activity and many other physiological functions show rhythmic patterns (Harker, 1964). Measurements are carried out, therefore, over at least a 24-hour period, as suggested by Phillipson (1962, 1963).

Since most predators, unlike herbivores (see III), can endure a period of hunger without harm, the duration of the experiment will not influence the results. On the other hand Grüm (1966) showed that with starved Carabidae activity increased after two days, whilst activity patterns changed after three days. Our experiments, however, are concluded before the activity of the animals increases exceptionally.

The motor activity pattern is often synchronized by light. The room where the apparatus stands is therefore illuminated according to the natural light regime of the appropriate season.

### d) Repetition of Measurements

PHILLIPSON (1962, 1963) and PHILLIPSON and WATSON (1965) demonstrated the variability of respiratory rate with size, physiological state (especially gonad condition) and season. Hence, measurements with adults or long-living developmental stages are repeated every three to five weeks throughout the year with newly captured animals at the appropriate temperature.

## 3. Calculation of Population Oxygen Consumption

Oxygen consumption is measured in the laboratory as cmm $O_2$ per individual per 24 hours. After weighing the animal on a Sartorius balance 2404, consumption is expressed as cmm $O_2/g/24$ h. Respiration of the population per unit of habitat (e. g. 1 ha) and per unit of time (1 year, as a rule) is then calculated, taking into account biomass lifetime (PETRUSEWICZ, 1967) of the different developmental stages at various temperatures as assessed from studies of the life cycle, population dynamics, and standing crop (FUNKE, chapter M; WEIDEMANN, chapter P). The "best estimate" (PHILLIPSON, 1963) of population oxygen consumption is obtained according to the equation (cmm $O_2/g \times 24$ h $\times$ average standing crop biomass $\times$ no. of days/ stage at a given temperature) *egg* + (cmm $O_2/g \times 24$ h $\times$ average standing crop biomass $\times$ no. of days/stage at a given temperature) *L1* + . . . . .

When species are studied that grow in relatively small steps over a large weight range, the regression line method of PHILLIPSON (1963) and WIEGERT (1965) may be used, i.e. estimation of oxygen consumption from the regression oxygen consumption/animal:body weight. This method presupposes either that temperature is constant over the whole period during which animals of a given weight range exist, or that the $Q_{10}$ remains constant for all weight classes within the possible temperature range. This is hardly to be expected according to findings in Carabidae (SCHMIDT, 1956), or other arthropods (see KEISTER and BUCK, 1964). Hence the regression line method may be employed only in special cases.

## 4. Computation of Population Energy Loss Due to Respiration

The "best estimate" of oxygen consumption will give quite different figures for respiratory energy loss according to whether fat, glycogen, or protein are assumed to be metabolized (PHILLIPSON, 1962). There are two possible ways of determining the energy resources used: a) measurement of the respiratory quotient $\left( RQ = \dfrac{CO_2}{O_2} \right)$, which will hardly give respresentative results, since $RQ$ varies with many factors (stage, nutritional state, activity, etc.) (WIGGLESWORTH, 1953); or b) chemical determination of resources: here, however, it is not possible to identify the component that is metabolized.

Population energy loss is therefore computed with either the oxycaloric equivalent of animal fat (4.72 kcal/l O$_2$) or an average calorific equivalent of 4.825 kcal/l O$_2$ after Brody (1945, cf. Wiegert, 1965), multiplied by the terms of the equation given above. After summing the energy loss of the population is expressed in kcal/ha/yr.

# IV. Assimilation

From the equation $E = A = P + R$ given in the introduction it follows that energy flow through a population can be evaluated by determination of total energy income, i.e. assimilation. Since potential energy is stored in the food eaten by the animals, assimilation is established by applying the following equations (after Phillipson, 1960):

a) weight of food eaten = weight of food offered — weight of food not eaten;

b) weight of food assimilated = weight of food eaten — weight of faeces.

When energy contents of food offered, food not eaten and faeces have been determined by calorimetry (see chapter I), energy assimilation is calculated from their respective energy contents according to the above equation.

If assimilation, or energy flow, can be determined in this way, it may be unnecessary to determine production and respiration. On the other hand the determination of assimilation can be corroborated by the results of independent measurements of production and respiration, and vice versa, i.e. if $A = P + R$ and $A = C - FU$ then $P + R = C - FU$ (for definitions see I and III 1f).

Since feeding biology in herbivores and predators differs widely, different methods have to be adopted for the assessment of assimilation.

## 1. Phytophagous Insects (Funke)

Feeding experiments to determine assimilation are performed mainly in laboratory cultures. Phyllophagous insects are fed on fresh beech foliage under nearly natural conditions (light regime, temperature). Extent of intake of food provided and food quantities can be determined by the weight differences between the food offered and the food not eaten. The energy content of the foliage relative to seasonal variation is determined by the research of Runge (chapter L).

A different method is used for animals eating smaller quantities over a period of time, and especially those which make holes or other patterns. Leaves are dried under pressure to flatten them. The eaten surface areas are outlined on graph paper for planimetry (Bray, 1961; other methods see Geyger, 1964). The thickness of each leaf is determined by means of a special instrument („Dickenmesser", Fa. Käfer, Meßuhrenfabrik, 722 Schwenningen/Neckar, W. Germany). Food consumption can be calculated from dry weights and energy contents of leaf squares of different thickness and for several seasons. The determination of food quantities would be faciliated if leaves of the same thickness could always be used.

The food quantities consumed in nature by animals eating well-defined patterns into leaves can be estimated for whole populations by sampling leaves from different levels in the canopy and different places in the study area and applying the data on leaves per m² of chapter B. In particular, this is done for all larval stages of leaf-miners. The energy content of the food eaten in these cases is calculated as follows: larvae are taken from the mines, exuviae and faeces are obtained by brushing or washing the leaves. Leaf parts of equal size and thickness (if possible, parts of the same leaf) are dried and the part with mines and the part without submitted separately for calorimetric measurement. Under the assumption already mentioned (III 1f), the energy content of the eaten foliage is obtained as the difference between the energy content of the samples.

Faeces of phyllophagous insects living on the beech are generally compact pellets; these are difficult to collect only for small animals and early larval stages.

For soil-dwelling larvae of phyllophagous beetles (see FUNKE, chapter M), no cultivation method is at present known, for an exact determination of food consumption and faeces.

## 2. Predatory Arthropods of the Ground Layer (WEIDEMANN)

### a) Food

Predators are rarely specialized with regard to their food. Hence, the kind of food taken in the natural habitat has to be determined. Groups which chew and swallow the food, such as centipedes, harvest spiders and most carabid beetles, have their gut content examined to show the spectrum of animals preyed upon. These investigations have to be repeated throughout that time of the year when the species in question is active, to make up a phenology of food preference. The material used for these investigations is taken from formalin traps (see WEIDEMANN, chapter P). The gut contents are embedded on a microscopic slide in Faure's solution (50 ml aqua dest., 30 g pulverized gum arabic, 20 ml glycerine, 50 g chloralhydrate, filtered through glass wool) and studied under the microcsope.

Food preferences of epigeic spiders may be detected by searching for food remains in nets (e. g. with Agelenidae), by feeding experiments with potential prey objects from the natural habitat (PHILLIPSON, 1960), or by the method of BREYMEYER (1967). She determined species composition and abundance of potential prey in 0.25 sqm cages from which spiders had been removed, in relation to undisturbed cages in the natural habitat. From the differences between the two types of cage after a definite time interval, it is possible to determine prey fed upon as well as the quantity of prey consumed by the enclosed spider population.

### b) Determination of Assimilated Food

Feeding experiments with representative prey objects are performed in the laboratory. Starved predators are given a weighed prey animal of known energy content. Food remains and faeces are weighed (dry weight) after an appropriate time interval, and their energy content determined by calorimetry. Since faeces are more or less liquid in centipedes, spiders and carabid beetles, and therefore can hardly be obtained quantitatively, assimilation cannot be determined exactly by feeding experiments in these groups.

## c) Determination of Food Quantity

The food quantity consumed during a definite time period may be determined by the Breymeyer method (see above).

Another method is suggested by studies of fish nutrition (Darnell, 1968). If the time for gut passage of a food is known from feeding experiments under nearly natural temperature conditions, food consumption in the habitat can be determined by evaluation of gut contents of individuals captured at intervals equal to the time of gut passage.

Lastly, the quantity of food consumed in the natural habitat may be evaluated from the growth rate of animals fed in the laboratory under nearly natural conditions on natural food (Darnell, 1968). From the data thus obtained and the biomass increment of the natural population, food consumption of the latter can be calculated.

## d) Evaluation of Assimilation (also for Phytophagous Insects)

From the results obtained by the methods mentioned, assimilation by the population can be calculated using the equations given above (p. 106) and taking into account the results of studies on population dynamics (see chapters M and P).

# References

Bray, J. R.: Measurement of leaf utilization as an index of minimum level of primary consumption. Oikos 12, 70—74 (1961).

Breymeyer, A.: Preliminary data for estimating the biological production of wandering spiders. (Secondary productivity of terrestrial ecosystems, pp. 821—834. Ed. K. Petrusewicz). Warszawa-Krakow: Polish Acad. Sci. 1967.

Darnell, R. M.: Animal nutrition in relation to secondary production. Amer. Zoologist 8, 83—93 (1968).

Geyger, E.: Methodische Untersuchungen zur Erfassung der assimilierenden Gesamtoberflächen von Wiesen. Ber. Geobot. Inst. ETH, Stiftg. Rübel, Zürich 35, 41—112 (1964).

Grüm, L.: Diurnal activity rhythm of starved Carabidae. Bull. Acad. Polon. Sci., Sect. Sci. biol. 14, 405—411 (1966).

Harker, J. E.: The physiology of diurnal rhythms. Cambridge: University Press 1964.

Keister, M., Buck, J.: Respiration: Some exogenous and endogenous effects on rate of respiration. In: The physiology of insects 3 (Ed. Rockstein, M.). New York, London: Academic Press 1964.

Lindemann, R. L.: The trophic-dynamic aspect of ecology. Ecology 23, 399—418 (1942).

Macfadyen, A.: Methods of investigation of productivity of invertebrates in terrestrial ecosystems. (Secondary productivity of terrestrial ecosystems, pp. 383—412. Ed. K. Petrusewicz). Warszawa-Krakow: Polish Acad. Sci. 1967.

Odum, E. P.: Fundamentals of ecology. 2. ed. Philadelphia-London: Saunders 1959.

— Energy flow in ecosystems: a historical review. Amer. Zoologist 8, 11—18 (1968).

Petrusewicz, K.: Concepts in studies on the secondary productivity of terrestrial ecosystem. (Secondary productivity of terrestrial ecosystems, pp. 17—49. Ed. K. Petrusewicz). Warszawa-Krakow: Polish Acad. Sci. 1967.

Phillipson, J.: The food consumption of different instars of *Mitopus morio* (F.) (Phalangiida) under natural conditions. J. Anim. Ecol. 29, 299—307 (1960).

— Respirometry and the study of energy turnover in natural systems with particular reference to harvest spiders (Phalangiida). Oikos 13, 311—322 (1962).

— The use of respiratory data in estimating annual respiratory metabolism, with particular reference to *Leiobunum rotundum* (Latr.) (Phalangiida). Oikos 14, 212—223 (1963).

PHILLIPSON, J.: A miniature bomb calorimeter for small biological samples. Oikos **15**, 130—139 (1964).
— WATSON, J.: Respiratory metabolism of the terrestrial isopod *Oniscus asellus* L. Oikos **16**, 78—87 (1965).
RICHMAN, S.: The transformation of energy by *Daphnia pulex*. Ecol. Monogr. **28**, 273—291 (1958).
SCHMIDT, G.: Der Stoffwechsel der Caraben (Ins. Coleopt.) und seine Beziehung zum Wasserhaushalt. Zool. Jb. Abt. Phys. **66**, 273—294 (1956).
WIEGERT, R. G.: Energy dynamics of the grasshopper populations in old field and alfalfa field ecosystems. Oikos **16**, 161—176 (1965).
WIGGLESWORTH, V. B.: The principles of insect physiology, 5th ed., London: Methuen & Co., New York: Dutton & Co. 1953.
ZINKLER, D.: Vergleichende Untersuchungen zur Atmungsphysiologie von Collembolen (Apterygota) und anderen Bodenkleinarthropoden. Z. Vergl. Physiol. **52**, 99—144 (1966).

# P. Food and Energy Turnover of Predatory Arthropods of the Soil Surface

Methods Used to Study Population Dynamics, Standing Crop and Production*

G. WEIDEMANN

## I. Introduction

In deciduous forests, such as the beech forests of the Solling, a large amount of leaf litter (ca. 3t/ha) accumulates every year and is the trophic base for a rich detritus food chain (ODUM, 1966). Within this food chain predatory arthropods, such as centipedes (Chilopoda), harvest spiders (Opiliones), web spiders (Araneae), ground beetles (Coleoptera, Carabidae), and rove beetles (Staphylinidae), inhabiting the litter and upper humus layer, are the main secondary (and tertiary) consumers. Since a great many of the phytophagous insects on beech spend part of their life cycles in the soil or on the soil surface (see FUNKE, chapter M), the epigeic predators are also interlinked with the grazing food chain (ODUM, 1966). It is therefore of great interest to study this group within the Solling Project in order to evaluate its contribution to the productivity of the whole community.

## II. Determination of Important Species in Relation to Productivity and Energy Flow

A great number of predatory species may be expected in a beech forest, so research economy requires that at the beginning of the study we should find out which species populations within a systematic group apparently contribute most to the productivity of the whole community at this trophic level.

The predatory arthropods of the soil surface are very mobile, with the exception of some spiders. Thus it is advisable to use pitfall traps for catching them.

Numbered honey jars, 12 cm deep, with narrow neck and 5.8 cm mouth opening, containing 4% formaline as preservative (after HEYDEMANN, 1956) and a detergent, are sunk into the ground. The trap is covered by a glass plate (15 × 15 cm) mounted on three wire feet as a protection against rain.

The traps are not distributed at random in the study area but situated singly or in groups at sites of different structure (with or without herbaceous layer, depressions etc.). This gives some insight into the dispersion of the species, too. The traps are changed at regular intervals throughout the year. Exchanged traps are fitted with a lid and then used as transport and storage vessels until the catch is sorted in the laboratory for further examination.

---

* For methods used to study energy flow see FUNKE and WEIDEMANN (chapter O).

Pitfall traps measure the "activity density" (Aktivitätsdichte, HEYDEMANN, 1953, 1956) of individuals per trap and per unit of time, but not the abundance (individuals/m²). From the trapping results therefore only the "activity dominance" (Aktivitäts-dominanz, HEYDEMANN, 1953) of each species can be calculated. Since the "activity density" of free-moving animals under otherwise identical conditions (e.g. structure of the habitat, weather) is correlated with size and motor activity, "activity dominance" is a good criterion to determine which species are presumably important in production ecology. On these species further research is focused.

## III. Life Cycle and Population Dynamics

It is impossible to determine the production of a population if the biology and the life cycle of the species in question are unknown. In the case of common species, the life cycle may be known in general from the literature, and one can begin to investigate the population dynamics under the special conditions of the study area. In other cases, the biology has to be elucidated first. This requires both field and laboratory work.

### 1. Phenology

The distribution in time of epigeic species is studied by pitfall trapping as described above. The traps are changed every fortnight for at least one year, and successive catches are represented as a phenological curve. Several curves may be drawn if different developmental stages (e.g. larvae) or age groups can be distinguished.

These curves offer much information: presumed breeding season(s), time of hatching from eggs, if first stages have been caught; duration of the stages, if successive stages have been captured; time of emergence from pupae in beetles, if not yet fully coloured and hardened individuals have been trapped. These findings must be verified, especially in regard to the breeding season. This is done by dissections of the gonads of females fixed in the preservative (formalin) in the traps. Breeding season is indicated by the presence of developing or ripe eggs in the ovaries.

### 2. Life Cycle

The full life cycle can often be traced only by bringing the respective species into the laboratory. This has been done with various carabid beetles and a spider. Carabids and spiders are kept in pairs or in greater numbers in refrigerator boxes or 1-l glass jars (see THIELE, 1968). The substratum is beech litter from the natural habitat or peat. Abiotic conditions (temperature, humidity, light regime) are kept as natural as possible. The beetles are fed twice a week with minced meat according to THIELE, and spiders with *Drosophila* or wax moths. Since the carabid larvae are cannibalistic, they are kept singly in suitable plastic cups.

From these cultures the missing data of the life cycles are obtained. The information may be tested in fenced field cultures established in the study area and checked at regular intervals.

A further task of the laboratory cultures is to find out which developmental stages, especially beetle larvae, belong to which adults.

## 3. Population Dynamics

From the definition of production given by MACFADYEN (1967) and PETRUSEWICZ (1967), which is followed in this paper (see section III 4), it follows that population dynamics and in particular the dynamic of abundance (Abundanzdynamik, SCHWERDT-FEGER, 1968) must be investigated quantitatively in order to determine production.

### a) Abundance of Eggs

The number of eggs per female can be determined in most cases only in the laboratory, especially in species where eggs are laid singly over a period of some weeks, as in Carabidae and Staphylinidae. In the other epigeic predators it will also hardly be possible to determine the abundance of eggs in the field. Thus, egg laying must be studied in laboratory cultures (see above).

When transferring the results to the natural population, it is necessary to know the sexual index during the breeding period. Sex ratio will be learned best from captures by hand (removal from random squares) since captures from pitfall traps may be influenced by the different activity of males and females during their reproductive phase (e.g. spiders). The abundance of eggs is then calculated as:

mean number of eggs per female × females per hectare.

### b) Abundance of Developmental Stages (Larvae) and Adults

The abundance of developmental stages and adults is determined by the quadrat method (BALOGH, 1958; MACFADYEN, 1963; SOUTHWOOD, 1966). For this purpose a sample plot 50 × 50 m is established in the study area. Posts are erected at its corners and every 5 m along its sides; a subdivision into 5 × 5 m squares is thus possible for orientation. 12 samples are taken every fortnight with an iron frame (25 × 25 cm) from the litter and humus layer according to random numbers previously drawn from a table of random numbers, as suggested by MACFADYEN (1963). The samples are brought into the laboratory in plastic bags and extracted by means of an extraction apparatus after KEMPSON et al. (1963). In the extractor, bowls of 30 cm diameter are used which can take half a sample. Three extractors, each with eight bowls, are operated simultaneously.

From the numbers of individuals obtained from each sample, the abundance and type of dispersion are calculated using the methods suggested by CONCELA DA FONSECA (1966). MACFADYEN (1963) and SOUTHWOOD (1967).

### c) Abundance of Large Carabidae

Since large arthropods often have an abundance too low to be detected by quadrat counting, the marking and recapture method is employed. According to SCHJØTZ-CHRISTENSEN (1965) an area of 100 m² was enclosed by means of 1 mm thick plastic foil ("Lamylux"), thus establishing an area closed to those species which are unable to fly. 41 1-l jars sunk into the ground were distributed at regular intervals over the plot; these traps are operated at different times during the year, corresponding to the phenological curves (see III1). They are checked every third day. The beetles so captured are taken into the laboratory and marked individually by means of a thermocauter. Marked beetles are released in the sample plot at the next check.

Abundance is calculated from the Lincoln index using the method of Schjøtz-Christensen for his closed locality.

The activity of the epigeic fauna is influenced by temperature and rainfall, as was demonstrated with Carabidae by Lauterbach (1964), for example. Under unfavourable weather conditions, as well as during the hibernation period or diapause, the Lincoln index method is not appropriate. In these cases, quadrat counting as described above is employed using $50 \times 50$ cm frames which also serve as controls for the results gained by marking and recapture.

## 4. Life Tables

From the abundances of the different developmental stages obtained by these methods, we are now able to construct life tables. Some difficulty arises from the overlapping of stages which occurs when the egg-laying period extends over a period longer than the duration of a stage. If the duration of the stage in question is known (e.g. from laboratory cultures under similar conditions), the total number of individuals that entered that stage may be calculated from the formula:

$$N = \frac{T}{t} \times \bar{n}$$

with $N =$ total number of individuals reaching a given stage;
$\quad \bar{n} =$ mean catch/sample date;
$\quad T =$ duration (days) of the period during which individuals of the stage in question have been captured, i.e. period between latest date without catch and latest date with catch;
$\quad t =$ duration of stage under natural conditions.

# IV. Standing Crop

"Standing crop is the quantitative status of a population on a specified area at a specified time moment" (Petrusewicz, 1967). Standing crop may be expressed in numbers (individuals/m²) which is the abundance (or summed abundances of different stages living together at the same time), or as biomass (g/m², live weight or dry weight), or as potential energy contained in the biomass (cal/m²).

For the determination of the standing crop biomass, a sufficient number of specimens are weighed every three to five weeks on a Sartorius balance 2404 sensitive to $\pm 0.01$ mg. Dry weight is determined after drying the material in a vacuum oven at 60° C for at least 24 hours. Biomass is calculated by multiplying the weight (wet or dry) by the density of the stage in question.

# V. Production

## 1. Total Production

### a) Production from Reproduction

Production from reproduction (*Pr*) of a population equals the abundance of reproducing females multiplied by the mean number of eggs per female and the mean egg weight. With predatory arthropods such data must be obtained in the laboratory, as mentioned above (III 3 a). In relation to *Pg*, *Pr* is almost negligible in these groups.

## b) Production from Individual Growth

Production from individual growth (*Pg*) includes both the increase in standing crop biomass (see IV) and the biomass of individuals that died during the defined interval. The latter is calculated by means of a survivorship curve which may be extracted from the life table (see III) as the difference in density between successive stages or with long-living stages (e. g. adults) between successive dates, multiplied by the mean weight of that stage during the corresponding period. The sum of these products plus the standing crop biomass remaining at the end of the period investigated constitute the population production due to individual growth. This relatively simple procedure can be used only if immigration is balanced by emigration, which may be assumed to be the case in the Solling beech forests under study, due to the size of the research area.

## 2. Minimum Production

Sometimes it may be impossible to carry out all the single measurements needed to compute the production of a population, either because the life cycle of a species is not fully known, or because it is too difficult to obtain exact data on the abundance of developmental stages. In such cases, some idea of the magnitude of production may be gained by determining minimum production: individuals that survive from the egg up to their first breeding season are the least number produced by the preceding generation, individuals that died during development not being counted. Therefore the standing crop of first-breeding individuals at the time of reproduction in species with a limited breeding season constitutes the minimum production of the preceding generation. This parameter seems to be valuable especially in species that reproduce more than once during their lifetime. When studying those species, e. g. Carabidae, as shown by Paarmann (1966) and Schjøtz-Christensen (1965), the proportion of first breeders has to be established by gonad dissection (Gilbert, 1956, Schjøtz-Christensen, 1965). Females that have passed at least one breeding period are marked by corpora lutea in the ovarioles.

The concept of minimum production in particular may be suitable for comparison of populations in different habitats or at different stages of succession. Minimum production has to be distinguished from production of imagines, as proposed by Funke (chapter M). The latter includes all individuals that survive to imaginal stage. Production of imagines will give a higher minimum production, for mortality in the adult stage has not yet influenced the population. Assuming that suitable methods are available for the determination of production of imagines, the latter parameter is better adapted for comparisons than is minimum production.

## VI. Preliminary Results

Pitfall trapping (12 traps) in a 120-year-old beech stand gave the following results with Carabidae in 1967/68:

| | No. | % |
|---|---|---|
| *Pterostichus oblongopunctatus* F. | 914 | 54.2 |
| *P. metallicus* F. | 280 | 16.6 |
| *Trechus quadristriatus* Schrk. | 354 | 20.9 |
| *Abax ater* Vill. | 22 | 1.3 |
| 5 other species | 116 | 7.0 |

An analysis of the captures of traps situated in different microhabitats (with or without herbaceous layer or depressions) showed *P. oblongopunctatus* and *P. metallicus* to be evenly dispersed in the plot (Fig. 1). Since the biology and ecology of *P. oblongopunctatus* were already known from the investigations of PAARMANN (1966), this species was chosen for further study.

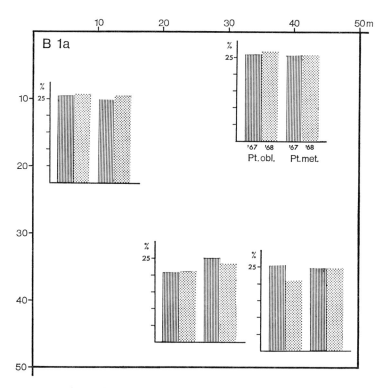

Fig. 1. Dispersion of *P. oblongopunctatus* and *P. metallicus* in the study plot B 1 a in 1967 and 1968 as indicated by trapping results. Each of the internal graphs represents the total captures of the respective species during one year gained with three pitfall traps

The phenology of *P. oblongopunctatus* in the study area as obtained by trapping is shown in Fig. 2.

The abundance of imagines was determined by the quadrat method in 1968. The data are summarized in Fig. 3 C 1.

The monthly biomass standing crop was calculated from mean live weight of males and females (Fig. 3 C 2) and the density figures (Fig. 3 C 1).

Gonad dissections in April 1968 showed that 50% of females had already passed at least one breeding season. Assuming the situation to be similar with males, since no criterion seems to exist to distinguish first-breeders in this sex, 50% of the standing crop of *P. oblongopunctatus* in the 1968 breeding season, i.e. 1,092 g/ha, would be the minimum production of the foregoing generation.

8*

This percentage of survivors from the previous generation seems to be very high in Carabidae. But in 1969 even more surprising results were obtained by a different method. Between May and August 1968, i.e. before hatching of adults of the new generation, 81 ♂ and 120 ♀ of *P. oblongopunctatus* were captured in the closed area

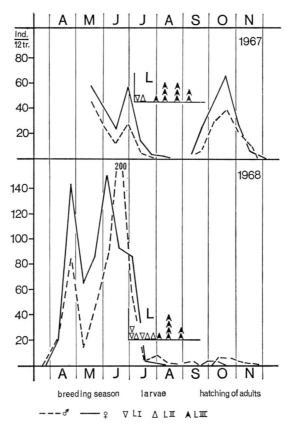

Fig. 2. Phenology of *P. oblongopunctatus* in the study plot B1a, 1967/68 as obtained by pitfall trapping

(see III 3 c) and marked individually. From these individuals 48 ♂ and 92 ♀ were recaptured during the following breeding season, April 29th to June 13th, that is 59.2% of the males and 76.6% of the females.

Oxygen consumption was measured (for methods see FUNKE and WEIDEMANN, 1970) in 1968 at temperatures near the monthly mean temperatures at 2 m above ground (Fig. 3A), since litter temperatures were not yet available. From the mean oxygen consumption per g per 24 hours (Fig. 3B), oxygen consumption of the natural population was calculated, taking into account the average biomass standing crop of imagines in the various months (Fig. 3D). The imaginal population of *P. oblongopunctatus* respired in 1968 1,700 l $O_2$/ha/year which equals an energy loss of 8,024 kcal/ha/year, if fat is assumed to be metabolized.

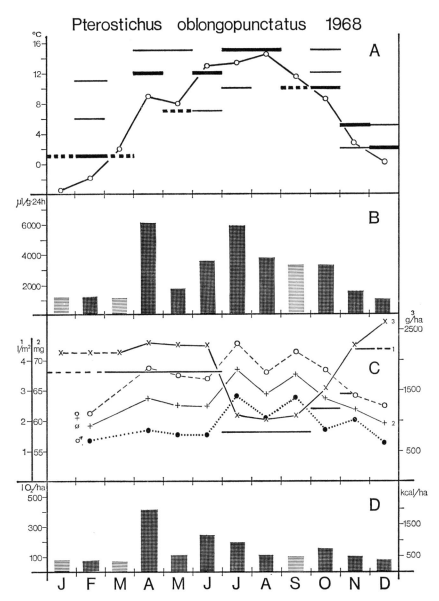

Fig. 3. **A**: Monthly mean temperature at 2 m above ground. Bars indicate the temperatures at which measurements of oxygen consumption have been made; heavy bars = temperatures on which calculations are based; broken bars — interpolation from foregoing or following month. **B**: Mean oxygen consumption in cmm/g × 24 h from measurements on single individuals over at least 24 hours. **C**: 1. Abundance (Ind./m²) as obtained by the quadrat method; broken line = interpolation; 2. mean live weight of ♂ (●) and ♀ (○) and average from both (+); 3. Monthly biomass standing crop as calculated from abundance and average weight assuming a sex ratio of 1:1. **D**: Energy loss due to respiration of the imaginal population as calculated from mean oxygen consumption and biomass standing crop. Fat was assumed to be metabolized (4.72 kcal/l O₂)

# References

BALOGH, J.: Lebensgemeinschaften der Landtiere, Berlin: Akademie-Verlag 1958.

CANCELA DA FONSECA, J.-P.: L'outil statistique en biologie du sol. III. Indices d'intérêt écologique. Rev. Ecol. Biol. Sol. **3**, 381—407 (1966).

GILBERT, O.: The natural histories of four species of *Calathus* (Coleoptera, Carabidae) living on sand dunes in Anglesey, North Wales. Oikos **7**, 22—47 (1956).

HEYDEMANN, B.: Agrarökologische Problematik. Diss. Kiel 1953.

— Über die Bedeutung der „Formalinfallen" für die zoologische Landesforschung. Faunist. Mitt. Norddeutschl. **1**, (6) 19—24, (1956).

KEMPSON, D., LLOYD, M., GHELARDI, R.: A new extractor for woodland litter. Pedobiologia **3**, 1—21 (1963).

LAUTERBACH, A.-W.: Verbreitungs- und aktivitätsbestimmende Faktoren bei Carabiden in sauerländischen Wäldern. Abh. Landesmus. Naturkunde Münster/Westf. **26** (4); (1964).

MACFADYEN, A.: Animal ecology, aims and methods, 2. ed., London: Pitman, 1963.

— Methods of investigation of productivity of invertebrates in terrestrial ecosystems. (Secondary productivity of terrestrial ecosystems, pp. 383—412. Ed. K. PETRUSEWICZ). Warszawa-Krakow: Polish Acad. Sci. 1967.

ODUM, E. P.: Ecology. New York: Holt, Rinehart & Winston 1966.

PAARMANN, W.: Vergleichende Untersuchungen über die Bindung zweier Carabidenarten (*P. angustatus* Dft. und *P. oblongopunctatus* F.) an ihre verschiedenen Lebensräume. Z. Wiss. Zool. **174**, 83—176 (1966).

PETRUSEWICZ, K.: Concepts in studies on the secondary productivity of terrestrial ecosystems. (Secondary productivity of terrestrial ecosystems, pp. 17—49. Ed. K. PETRUSEWICZ). Warszawa-Krakow: Polish Acad. Sci. 1967.

SCHJØTZ-CHRISTENSEN, B.: Biology and population studies of Carabidae of the Corynephoretum. Natura Jutlandica, Aarhus **11**, 1—173 + 1—72 (tables) 1965.

SCHWERDTFEGER, F.: Ökologie der Tiere II: Demökologie, Hamburg-Berlin: Parey, 1968.

SOUTHWOOD, T. R. E.: Ecological methods. London: Methuen, 1967.

THIELE, H.-U.: Zur Methode der Laboratoriumszucht von Carabiden. Decheniana **120**, 335—342 (1968).

# Q. On the Abundance of Bacteria and Other Microorganisms

G. Niese

## I. Introduction

In connection with the IBP Solling Project, the occurrence of microorganisms which are predominant in the mineralization of the organic matter synthesized by autotrophic plants and partly altered in composition by animals was examined in the soils of the Solling area. The investigations were carried out from April 1967 until March 1968 by G. Jagnow and continued by the author.

After preliminary investigation of a series of problems relevant to the project in question, enumerations of the microorganisms listed below were carried out:

1. Bacteria and bacterial spores on soil extract glucose agar,
2. Fungi on malt extract agar,
3. Actinomycetes on rose bengal malt extract agar,
4. Nitrite- and nitrate-forming microorganisms in a liquid medium containing $(NH_3)_2SO_4$ and mineral salts,
5. *Azotobacter chroococcum*,
6. Anaerobic spore-forming bacteria (Clostridia) in a substrate free of N.

## II. Methods and Material

### 1. Sampling of the Soil

Using a bore-stick ($\varnothing * 30$ mm), 25 separate samples were taken every 4 to 6 weeks always from the same plot. These samples were thoroughly mixed in order to obtain one sample of 600 to 800 g of soil. After brief cooling in a refrigerator, the samples were transported for about 7 hours in an insulated bag together with 6 commercial cooling cartridges. After the soil had been stored overnight at $+4°$ C, the investigations were started the next morning. Water content and pH were determined for each sample.

Samples were taken from the following sites of the Solling area:

| | | | |
|---|---|---|---|
| A | Beech stand B1 | Humusstoff-horizon (except litter) | |
| B | Beech stand B1 | mineral soil | 0—15 cm |
| C | Grassland W2 | NPK-plot | 0— 5 cm |
| D | Grassland W2 | NPK-plot | 5—15 cm |
| E | Grassland W2 | NPK-plot | 15—25 cm |
| F | Grassland W2 | unfertilized plot | 0— 5 cm |
| G | Grassland W2 | unfertilized plot | 5—15 cm |
| H | Grassland W2 | unfertilized plot | 15—25 cm |

* $\varnothing$ means diameter.

## 2. Preparing the Soil Suspensions for the Plate Counts

50 g of fresh soil were shaken with 450 ml 0.18% Na-pyrophosphate solution for 30 min. After diluting the soil suspensions in the usual way with sterile tap water, the solid and liquid media were inoculated and incubated at 25° C.

## 3. Solid Media Used for the Plate Counts

### a) Soil Extract Glucose Agar

According to Spicher (1952/54), modified by Singh-Verma (1964): Glucose 1.5 g, $FeSO_4 \cdot 7 H_2O$ 0.05 g, $Na_2MO_4 \cdot 2 H_2O$ 0.05 g, $K_2HPO_4$ 1.0 g, $MgSO_4 \cdot 7 H_2O$ 0.5 g, $CaCO_3$ 1.0 g, Difco yeast extract 0.10 g, Hoagland solution, conc. 1% 10.00 ml, soil extract 50.0 ml, dist. water 940.0 ml, agar-agar 15.0 g; pH 7.0. Sterilization: The constituents, except the glucose, were dissolved in 800 ml of water, filled in small portions into Erlenmeyer flasks and sterilized at 121° C. The glucose was dissolved separately in 200 ml of water and sterilized by filtration. Before use, an equivalent volume of the slightly heated solution of glucose was mixed with the liquefied agar medium.

### b) Malt Extract Agar

Biomalz (commercial malt extract) 20.0 g, agar-agar 20.0 g, dist. water 1000.0 ml; pH 5.0. Sterilization: 10 min at 121° C.

### c) Rose Bengal Malt Extract Agar

(Ottow and Glathe 1968): Biomalz (commercial malt extract) 20.0 g, rose bengal 0.067 g, $KH_2PO_4$ 0.5 g, Hoagland solution, conc. 1% 10.0 ml, soil extract 50.0 ml, agar-agar 18.0 g, tap water 940.0 ml; pH 6.0—6.2. Sterilization: 10 min at 121° C.

## 4. Liquid Media Used for Tube Counts

### a) Nitrite- and Nitrate-forming Bacteria

$K_2HPO_4$ 1.0 g, $FeSO_4 \cdot 7 H_2O$ 0.03 g, NaCl 0.30 g, $MgSO_4 \cdot 7 H_2O$ 0.30 g, $CaCO_2$ 7.50 g, dist. water 1000.0 ml. 25 ml of medium were pipetted into 100 ml Erlenmeyer flasks and sterilized for 10 min at 121° C. Just before inoculation 1 ml of a solution containing 2% $(NH_4)_2SO_4$, sterilized by filtration, was added. After 4 weeks of incubation at 25° C the flasks were tested for the presence of nitrite and nitrate according to the method published by Pramer and Schmidt (1964).

### b) $N_2$-fixing Anaerobic Spore-forming Bacteria (Clostridia)

Glucose 20.0 g, $K_2HPO_4$ 1.0 g, $MgSO_4 \cdot 7 H_2O$ 0.20 g, NaCl 0.01 g, $MnSO_4 \cdot 4 H_2O$ 0.01 g, $CaCO_3$ 30.0 g, dist. water 1000.0 ml. Tubes of 16 mm diameter, containing 1 Durham tube, were filled with 10 ml of solution and sterilized for 10 min at 121° C. After inoculation the solution was sealed up by 2 ml liquefied rotted water agar (3%). After 2 weeks incubation at 25° C those tubes showing gas formation were regarded as positive for the development of anaerobic spore-forming bacteria, capable of fixing $N_2$.

Using liquid media, the most probable numbers of microbes were calculated according to the tables of Mc Crady (see Pochow and Tardieux, 1962).

## III. Preliminary Results

Since 1967 investigations of the microflora in the forest and grassland soils of the Solling area have been carried out, the results revealed considerable variation in the numbers of the soil microorganisms.

In contrast to the numbers of bacteria, those of the bacterial spores in the Humusstoff-horizon of the soil of the beech stand B1 showed some correlation with the seasons, decreasing during 1967 until autumn. During the winter of 1967/68 the content of bacterial spores rose strongly, decreasing again during 1968. As the humidity of the soil did not vary much, the changes in the numbers of bacterial spores may have been caused mainly by temperature variations in the upper layers of the soil. Rising temperatures in summer improved the conditions for the development of active cells; on the contrary, with decreasing soil temperatures in autumn and winter the conditions for microbial growth deteriorated. Therefore, most of the soil bacteria formed inactive spores.

In the mineral soil the numbers of the bacterial spores were not related to the seasons. Maximum and minimum values were found at various times during the year.

The numbers of fungi and actinomycetes in the soil of the beech plot B1 were not clearly related to the seasons, either. There was an especially rapid fall in the content of fungi and actinomycetes in the Humusstoff-horizon during June and July 1968. In the following months the numbers of fungi rose while those of the actinomycetes decreased again after having risen during a short period.

In the mineral soil of the beech stand B1, the numbers of fungi decreased considerably from August to the end of 1967, but later they multiplied again. During summer and autumn 1968 the contents of fungi and actinomycetes remained mostly at the same level.

It may be supposed that the content of bacteria as well as the contents of fungi and actinomycetes in the soil of the grassland W2 depend to some extent on the seasons. In the upper layers of the grassland soil (0—5 cm) the numbers of bacteria did not change very much throughout the year, but in summer and in early autumn low numbers of bacteria were found in deeper layers; later on the bacteria increased again. The numbers of fungi and actinomycetes, too, were low in summer, but these organisms developed less distinctly in autumn than did the bacteria.

The NPK fertilization of the grassland did not essentially influence its content of microorganisms. A slight increase in the numbers of bacteria, fungi and actinomycetes occurred only in the upper horizons of the NPK plot.

In the soil of the beech stand, only a very small number of nitrite-forming microorganisms and no nitrate-forming organisms could be detected. In the grassland soil, however, nitrite- and nitrate-forming microorganisms were found to be present during the whole of 1968. The fertilization of the grassland resulted in an increase in the numbers of nitrate-forming microbes.

As the development of nitrifying bacteria is inhibited in acid soils, the relatively small numbers of these organisms are in accordance with the low pH values of the soils investigated.

The acid soils of the beech stand B1 and grassland W2 also contained very few anaerobic spore-forming bacteria capable of fixing $N_2$, and no cells of *Azotobacter chroococcum*. Therefore the assimilation of gaseous nitrogen by free-living microbes in acid forest and grassland soils cannot be regarded as a significant factor in the N supply of these soils.

Additional conclusions may be drawn from the results of the microbiological investigations when more detailed information is available about the other research projects being carried out in the Solling area.

# References

Ottow, J. C. G., Glathe, H.: Rose bengal-malt extract-agar, a simple medium for the simultaneous isolation and enumeration of fungi and actinomycetes from soil. Appl. Microbiol. **16**, 170—171 (1968).

Pochon, J., Tardieux, P.: Techniques d'analyse en microbiologie du sol. St. Mande, Seine, Ed. de la Tourelle 1962.

Pramer, D., Schmidt, E. L.: Experimental Soil Microbiology. Minneapolis: Burgess Publishing Co., 1964.

Singh Verma, S. B.: Zum Problem des quantitativen Nachweises der Mikroflora des Bodens mit der Methode Koch. Diss. Landw. Fak. Univ. Gießen, 1964.

Spicher, G.: Untersuchungen über die Wechselwirkungen zwischen *Azotobacter chroococcum* und höheren Pflanzen. Zbl. Bakt. II **107**, 356—378 (1952/54).

# R. Microbial Transformation of Organic Material in the Soil

J. Gnittke, Ch. Kunze and L. Steubing

## I. Introduction

Every year a large amount of organic material is added to the soil by plant litter. This material serves as a source of food and energy for heterotrophic soil microorganisms. The most abundant organic constituent of plants is cellulose, which comprises 15—60% of the dry weight. The content of hemicelluloses ranges from 10—30% and of lignin from about 5—30%. With the decomposition of the organic material, the amount of lignin in the soil increases while the cellulose declines (Beck, 1968). There is a smaller water-soluble soil fraction comprising amino acids, aliphatic acids, simple sugars and tannins, and contributing 5 to 30% of the tissue weight. The ether- and alcohol-soluble fraction contains a number of pigments, resins, waxes, fats and oils. Finally, there is a protein fraction which contains a large amount of organic nitrogen and sulfur (Alexander, 1967). The studies of our working group concerned the presence and decomposition of cellulose, lignin, amino acids, tannins and some pigments in the soil of beech and spruce plots in the Solling Project.

## II. Methods and Some Results

### 1. Cellulose

The content of cellulose in soil samples was determined by the method of Black (1950).

Table 1. *Annual average content of cellulose in different horizons of beech and spruce soils*

| Horizon | Cellulose % | | Hexoses % | | Nitrogen % | |
|---------|-------------|--------|-----------|--------|------------|--------|
|         | Beech | Spruce | Beech | Spruce | Beech | Spruce |
| F       | 19    | 21     |       |        |       |        |
| H       | 13    | 14     | 19.5  | 63.8   | 0.65  | 0.5    |
| A₀      | 2     | 2      | 2.4   | 3.0    | 0.1   | 0.1    |

The amount of cellulose declines rapidly in the Solling soil under beech and spruce stands. There was no significant correlation between the amount of this carbohydrate in different soil horizons and the season, but there was a statistically significant excess in the annual average of cellulose in the spruce soil compared with the beech (Steubing, 1970b).

Measurements on cellulose decomposition under environmental conditions followed the method of Unger (1960): Nylon net bags containing cotton as a cellulose model were placed in different layers of the soil. After 10 weeks the loss of cotton dry weight was determined.

Investigations of the decomposition of cotton showed that the microbial attack on this material was stronger in the beech soil. It is well known that a higher level of of nitrogen promotes the destruction of cellulose while a higher content of carbohydrates retards it. Table 1 shows that better conditions for cellulose breakdown existed in the beech soil.

Table 2. *Decomposition of different cellulose-containing materials in the beech soil*

| Time | Substance | Decomposition in % of the dry weight of the cellulose-containing material in horizon | | |
| | | F | H | A$_e$ |
| --- | --- | --- | --- | --- |
| 1. 5.—1. 8. | cotton | 40 | 38 | 23 |
| | roots | 40 | 52 | 51 |
| | oat straw | 34 | 24 | 32 |
| 4. 6.—2. 9. | cotton | 40 | 49 | 46 |
| | roots | 71 | 65 | 61 |
| | oat straw | 22 | 40 | 27 |
| 1. 7.—1. 10. | cotton | 35 | 53 | 44 |
| | roots | 62 | 76 | 77 |
| | oat straw | 30 | 35 | 25 |
| 1. 8.—1. 11. | cotton | 10 | 26 | 33 |
| | roots | 47 | 68 | 55 |
| | oat straw | 24 | 20 | 19 |

When cotton, oat straw and roots of *Taraxacum officinale* were buried in the soil, analyses gave the highest cellulose loss for roots, followed by cotton, then incubated straw (Table 2).

## 2. Lignin

Lignin determinations in beech and spruce soils over the whole year yielded more lignin under the spruce than under the beech stand (Table 3).

Table 3. *Annual average lignin content of the soil dry weight*

| Horizon | Beech soil | Spruce soil |
| --- | --- | --- |
| F | 44% | 48% |
| H | 36% | 48% |
| A$_e$ | 4% | 3% |

Besides these lignin analyses in the soil samples, there were investigations into the decomposition of vanillin. Vanillin is a low-molecular-weight breakdown product of lignin. It was found as a native product in our soil samples. A total of seven low-molecular-weight phenols were isolated: p-hydroxybenzoic acid, protocatechuic acid, vanillic acid, ferulic acid, p-coumaric acid and two aldehydes: vanillin and p-hydroxy-benzoic aldehyde. Fig. 1 shows the UV-absorption spectra of the vanillin and p-hydroxybenzoic aldehyde isolated from soil samples. The curves of the isolates are

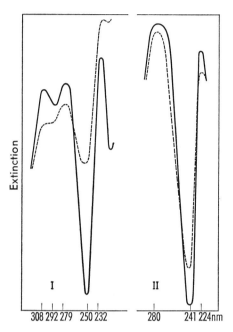

Fig. 1. UV-absorption spectrum of vanillin (I) and p-hydroxybenzoic aldehyde (II).
———— test substance; – – – – isolated substance from the soil

identical with the spectra of the standard substances. The calculation of the quantity of the seven phenols in Solling soils gives a concentration of 3—7 kg/ha to 10 cm soil depth (KUNZE, 1969a). Relating the phenol concentration to the organic substance of the soil gave a value of 0.002—0.005%. The estimations of WHITEHEAD (1964) and FLAIG (1967) gave similar results for different soils.

In laboratory experiments, we added 0.1% vanillin to the soil samples. The decomposition of this phenolic compound was studied. At the same time the respiration of soil samples was determined, so the two different kinds of soil activity could be compared. We found the maximum activity for both during the period October to December. The results of analyses made every 3 weeks are calculated for 2 months. Figs. 2 and 3 show the results. It is very important to see that the values and therefore the relative levels of soil activity in the different horizons change with the parameter, taking soil weight, the highest microbial activity was found in the F horizon; taking volume, the highest level was calculated for the H horizon and taking carbon content, soil respiration and vanillin decomposition were greatest in the $A_e$ layer (Kunze, 1969c).

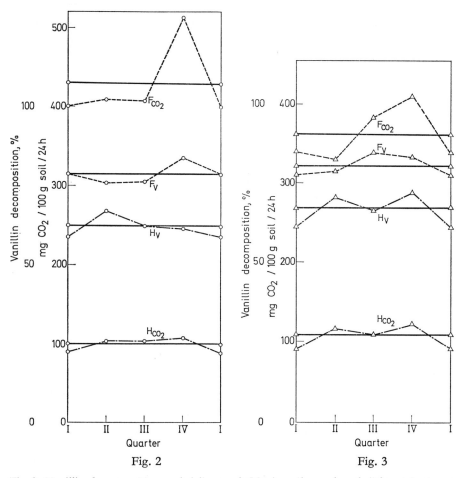

Fig. 2

Fig. 3

Fig. 2. Vanillin decomposition and delivery of $CO_2$ in soil samples of different horizons under beech stands

Fig. 3. Vanillin decomposition and delivery of $CO_2$ in soil samples of different horizons under spruce stands

## 3. Amino Acids

If the quantity of amino acids in the F-horizon is fixed at 100, there was a rapid decline to 50 in the H layer and there were only traces at 1 m depth in the Solling soil (Fig. 4). Soil amino acids originate in microorganisms, plant roots or dead animals, but diaminopimelic acid (DAP) is a protein of the bacterial cell wall; we therefore tried to estimate the mass of bacteria in the soil profile by chemical analyses of DAP concentration. A technique for isolating DAP from soil samples (Steubing, 1970a) enables us to correlate the amount of this amino acid with the number of bacteria in the different horizons (plate-count method). The decline in DAP concentration with increasing soil depth seems to follow the same trend as the mass of bacteria. So far, no exact information can be given about the level of agreement between the values compared. There is no doubt that an advantage of the DAP method is its good reproducibility, whereas the determination of the whole germs of bacteria is problematical.

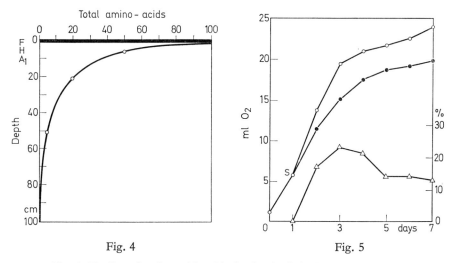

Fig. 4                                          Fig. 5

Fig. 4. Decline of amino acids with the depth of the beech soil ($A_1 \triangleq A_e$)

Fig. 5. The effect of adding streptomycin to soil samples upon the activity of catalase.
o—o without and ●—● with streptomycin (100 $\gamma$/ml) △—△ retarding effect in %;
S = time of adding streptomycin

## 4. Enzymes

We have investigated the possibility of obtaining information about microbial activity by studying soil enzymes. For this we chose catalase, a widely distributed enzyme, which catalyses the following reaction:

$$2 H_2O_2 \to 2 H_2O + O_2.$$

First results show that the activity of catalase is significantly inhibited by the addition of streptomycin to the soil samples (Fig. 5), from which we may conclude that soil catalase has a microbial origin (KUNZE, 1969b). Further investigations will show whether there is a correlation between the mass of living microorganisms and the activity of catalase.

## 5. Tannins

The vast and chemically heterogeneous group of the tannins was represented by the decomposition of tannin (DAB 6). In laboratory experiments this substance was added in different concentrations to soil samples, and its breakdown was studied by measuring the release of $CO_2$ from the samples. In another series the activity of catalase was examined in relation to the tannin concentration. Determination of the tannins was effected by a method similar to that of Constabel (1968) and Friedrich (1958). At the beginning of the test a retarding effect was seen with increasing

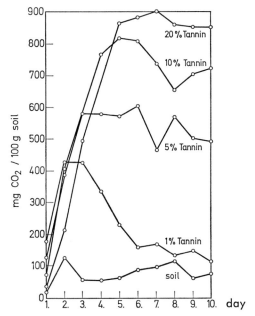

Fig. 6. Delivery of $CO_2$ depending on the time

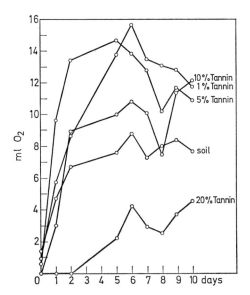

Fig. 7. Activity of catalase depending on the time

concentration of tannins, but at the end soil respiration was higher for all the tannin-soil mixtures than for the untreated control, as Fig. 6 shows. It demonstrates that there is a remarkable effect not only by the quantity of tannin but also by the time: as the investigation proceeds, maximum $CO_2$ production changes from the lower to higher tannin concentrations.

The activity of catalase increased with the addition of tannin concentrations up to 5% (Fig. 7); larger amounts retarded it for a limited period. This delaying effect became weaker and was transformed into an acceleration after 5 days for a 10% tannin concentration, incubated in the soil.

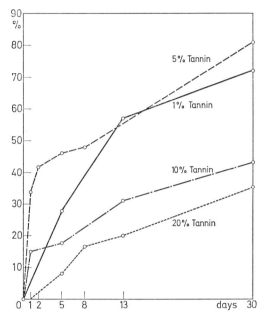

Fig. 8. Microbial breakdown of different concentrated tannin-soil mixtures depending on the time

In experiments on the breakdown of tannins one can also see a reflection of the increasing soil-tannin concentration in the biological activity. Tannin concentrations of 1 to 5% are decomposed almost three times as fast (in 30 days) as those of 10 and 20% (Fig. 8).

## 6. Chlorophyll Content

Every soil contains chlorophyll. GORHAM and SANGER (1967) and CORNFORTH (1969) extracted the pigment from woodland soils and found a relation between the amount of chlorophyll and primary production. Using the method of HOYT (1967), the chlorophyll units in our soil samples were determined. In very acid substrate (pH 3.5—4), there is destruction of chlorophyll with formation of phaeophytin. Therefore we supposed that the chlorophyll found in our samples has its origin, not only in the above-ground plant cover, but also in the autotrophic soil algae. So we sampled at intervals of one month the quantity of chlorophyll and of algae (fluorescent microscopy, following Tchan, 1952) in the different horizons (Table 4). There was the same declining trend from the litter to the mineral horizon. Yet there was also a gradual increase in chlorophyll in the F horizon not matched by the increase in algae. Thus, the chlorophyll in the F-layer must proceed not only from the soil algae but also from fresh remains of green leaves or in some cases from the excrement of herbivores (Steubing, 1970a).

We need further investigations to confirm these results. Preliminary explorations are going on to learn more about the blue-green algae by extracting phycocyan from the soil and analyzing it.

Table 4. *Chlorophyll content and numbers of algae from different soil horizons*

| Time | Horizon | Units of chlorophyll/100 g | | Algae/1 g | |
|------|---------|---------------------------|---|-----------|---|
| 1. 3. 1969 | F | 1.8 | | 149 000 | |
| | | | *** | | *** |
| | H | 0.39 | | 83 290 | |
| | | | * | | n. s. |
| | A$_\theta$ | 0.33 | | 94 730 | |
| 1. 4. 1969 | F | 15.9 | | 205 200 | |
| | | | *** | | *** |
| | H | 23.6 | | 270 190 | |
| | | | * | | ** |
| | A$_\theta$ | 4.1 | | 24 630 | |

# References

Alexander, F. E., Jackson, R. M.: Examination of soil microorganisms in their natural environment. Nature (Lond.) 174, 750—751 (1954).

Beck, Th.: Mikrobiologie des Bodens. München-Basel-Wien: Bayer. Landwirtschaftsverlag, 1968.

Black, W. A. P.: The seasonal variation in the cellulose content of the common scottish Laminariaceae and Fucaceae. J. Mar. Biol. Ass. U. Kingd. 29, 379—387 (1950).

Constabel, F.: Gerbstoffproduktion der Calluskulturen von Juniperus communis. Planta (Berl.) 79, 58—65 (1968).

Cornforth, J. S.: Chlorophyll compounds and nitrogen availability in Indian soils. Plant and Soil. 30, 113—117 (1969).

Flaig, W.: Zersetzung und Humifizierung. Vortrag zur Sitzung der Kommission II, IV und V der Deutschen Bodenkundlichen Gesellschaft 1967.

Friedrich, H.: Untersuchungen über die phenolischen Inhaltsstoffe von Prunus communis L. Pharmazie 13, 299—305 (1958).

Gorham, E., Sanger, J.: Plant pigments in woodland soils. Ecology 48, 306—308 (1967).

Hoyt, P. B.: Chlorophyll-type compounds in soil. I. Their origin. Plant and Soil 25, 167—180 (1966).

— Chlorophyll-type compounds in soil. II. Their decomposition. Plant and Soil 25, 313—328 (1966).

— Chlorophyll-type compounds in soil. III. Their significance in arable soils. Plant and Soil 26, 5—13 (1967).

KUNZE, Ch.: Isolierung und Identifizierung phenolischer Ligninspaltstücke aus Waldböden. Plant and Soil **31**, 389—390 (1969a).
— Der Einfluß von Streptomycin und aromatischen Carbonsäuren auf die Katalase-Aktivität in Bodenproben. Zbl. Bakt. II (in press), 1969b.
— Die biologische Aktivität von Bodenproben, gemessen an der $CO_2$-Abgabe und dem Vanillinabbau. Angew. Botan. **43**, 149—157 (1969c).
STEUBING, L.: Chemische Methoden zur Bewertung des mengenmäßigen Vorkommens von Bakterien und Algen im Boden. Zbl. Bakt. **124**, 245—249 (1970a).
— Soil flora: Studies of the number and activity of microorganisms in woodland soils. In: Analysis of temperate forest ecosystems. (D. E. REICHLE, ed.) Berlin-Heidelberg-New York: Springer 1970b.
TCHAN, Y. T.: Counting soil algae by direct fluorescence microscopy. Nature (Lond.) **170**, 328—329 (1952).
UNGER, H.: Der Zellulosetest, eine Methode zur Ermittlung der zellulolytischen Aktivität des Bodens in Freilandversuchen. Z. Pflanzenernähr. Düngung., Bodenkunde **91**, 44—52 (1960).
WHITEHEAD, D. C.: Identification of p-Hydroxybenzoic-, Vanillic-, p-Coumaric- and Ferulic Acids in Soils. Nature (Lond.) **202/1**, 417—418 (1964).

# S. The Measurement of Climatic Elements which Determine Production in Various Plant Stands

## Methods and Preliminary Results

### O. KIESE

## I. The Task

According to BAUMGARTNER (1963), the plant absorbs approximately 1% of the short-wave radiant energy available per year, and contributes up to 5% to the nocturnal heat balance by respiration. According to research by GALOUX (1968) the consumption of photosynthetically effective energy can amount to 1.8% of the available energy related to the vegetation period.

The task imposed by the Solling Project upon the meteorological investigators is to trace the course of such diverse climatic elements as soil and air temperature, air humidity, wind velocity etc., and to determine the separate components of the heat balance (radiation, flow of latent and sensible heat, flow of soil heat, heat flow in the mass of the stand) in order to help to analyse the causal relationships between the environmental factors and the physiological production of plants and animals.

The precipitation measurements are being carried out in connection with research on the water balance by the "Institut für Bodenkunde und Waldernährung" of the University of Göttingen (BENECKE and MAYER). The close connection between heat and water balance through evaporation means that there is especially close cooperation between the two research groups as regards the Solling Project.

## II. The Measuring Stations

Two different plant stands growing on sites of approximately the same altitude were chosen for intensive microclimatic research: a meadow (sample area W 1) and a stand of old beech trees (sample area B 1).

### 1. The Measuring Field on W 1

The sample area W 1 (Intr., Fig. 1) is situated on a slope with an inclination of approx. 3° to the north. A meadow, part of which is used as pasture, adjoins W 1 on the west. The meteorological measuring field covers an area of $20 \times 21$ m² in the centre of the meadow land, so that the same vegetation is found for at least 80 m all around the measuring field.

To the east is a plantation of spruce *(Picea abies)*, 1 m high and approximately 30 m wide, which is itself bounded by a raised bog where peat is no longer cut and where a variety of trees of different heights are now growing. To the south the boundary is formed by a strip of wild shrubs and trees, beyond which is a cultivated area divided into narrow plots. The plots are utilized as pasture land, for spruce plantations and, further to the south-west, for agriculture. To the north-west and the north, W 1 inclines towards a flat valley-like depression some 80 m wide, with young spruce trees growing in the bottom. This nursery is separated from a tall spruce forest by a road.

The inclination of the slope and the tree stands restrict the horizon, producing an interference which is especially noticeable in the segment of the azimuth from south-east to west. To the south the sky is partly concealed by trees to an angle of 14°. The measuring field itself shows small undulations of the surface parallel to the slope (h < 0.3 m). On the whole the situation is not ideal, but it does represent an open unconfined area typical of the Solling at this altitude (∼ 480 m above sea level).

## 2. The Measuring Field on B 1

The measuring field on plot B 1 (Intr., Fig. 5) takes up an area of $10 \times 20$ m² within a very large beech *(Fagus silvatica)* stand. It is 500 m above sea level and almost horizontal (inclination < 2% to the south), just below one of those flat-topped hills (505 m) typical of the Solling. To the north there is a mature spruce forest (F 1) some 400 m distant, to the west the beech forest gives way after about 200 m to a plantation of young spruce trees some 6 to 10 m high. To the east and the south the beech forest extends for about 1 km, after which there are spruce stands. Being in the middle of a forest area, the site may be considered as closed on all sides. This factor is of importance, particularly as regards the penetration of wind into the stand.

The timber stand of B 1 consists of 120-year-old beech trees (Intr., Fig. 3). The majority of the trees possess smooth trunks up to 10 m. At this level the branching begins. The average height of the trees is up to 26 m, some specimens being about 28 m. The canopy is not completely closed but is so dense that the development of undergrowth is nearly prevented. This means that plot B 1 shows a distinct vertical division:  0—10 m = trunks,

> 10—25 m = crowns,
> 25—28 m = tops.

## 3. Other Measuring Fields

Besides plots W 1 and B 1, the areas F 1 (spruce stand) and B 4 (young beech stand) are included in the measuring program, too (Intr., Fig. 1). On each of the areas mentioned there are a conventional meteorological screen equipped with a thermo-hygrograph, a screen psychrometer, and maximum-minimum thermometers. Thus comparable climatic data are obtained from a considerable area of the Solling.

On B 4 there is in addition a small meteorological screen (Israel-screen) 12 m high containing only a thermohygrograph, to provide roughly comparable data for B 4.

## III. Recording the Data

An economical method of recording measured values is essential for constant assessment of the components of the heat balance. For this purpose the Siemens Company, Karlsruhe, have developed novel data-processing equipment for the meteorological measurements in cooperation with A. BAUMGARTNER, München.

This system permits all sensors to be scanned at very short intervals of time and also converts and stores the measured values.

The measuring process is as follows: a measuring point selector connects all measuring points in turn via an amplifier with the input of a digital voltmeter. The output values of the voltmeter are fed into a small computer (Siemens 101), coordination being controlled electronically from the measuring point selector. Measuring cycle intervals and measuring periods can be set on the computer control panel. During 1969 the measuring period was 12 min and the cycle interval 20 sec; i.e. each measuring point was scanned 36 times per 12-min period.

The measured values are stored and at the end of each measuring period printed out by a teleprinter in figures and simultaneously coded on paper tape. It is possible to transmit via the control panel to each measuring point conversion factors (4-figure summands, 4-figure factors and powers of ten) which will ensure that the data print-out has the correct sign, decimal place and dimension. Each printout starts with a code number representing the date and time of day (see Table 1).

Table 1. *Part of a data print out (reduced)*

| | 2471744 | | | | | | |
|---|---|---|---|---|---|---|---|
| 35 | | | | | | | |
| 1+ 11.95 | 2+ 9.100 | 3+ 12.05 | 4+ 9.350 | 5+ 12.15 | 6+ 9.250 | 7+ 11.80 | 8+ 9.300 |
| 9+ 11.70 | 10+ 9.600 | 11+ 12.75 | 12+ 12.80 | 13+ 12.55 | 14+ 12.75 | 15+ 11.30 | 16+ 11.60 |
| 17+ 44.30 | 18+ 48.45 | 19+ 10.80 | 20+ 11.05 | 21+ 10.65 | 22+ 12.20 | 23+ 11.95 | 24+ 12.00 |
| 25+ 0.170 | 26+ 0.790 | 27+ 2.090 | 28+ 8.500 | 29+ 8.170 | 30+ 10.27 | 31+ 11.30 | 32+ 0.200 |
| 33+ 75.00 | 34+ 43.05 | 35- 0.100 | 36+ 1.120 | 37+ 0.100 | 38+ 0.090 | 39+ 0.000 | 40+ 0.000 |
| 41+ 1.934 | 42+ 1.497 | 43+ 1.233 | 44+ 0.827 | 45+ 0.188 | 46+ 0.163 | 47+ 0.261 | 48+ 0.294 |
| 49+ 0.188 | | | | | | | |

It takes 3 min to print out the data from 60 measuring points, so that each measuring period is necessarily followed by a 3-min pause. A pre-programmed computation step averages the stored data by dividing the sum by the number of measuring cycles. By feeding in suitable factors, the same computation can be made to yield the sum for the period in question, when required, as it is in radiation measurement.

In addition to storing measured values, the recorder system can be set to scan individual values from outside or to scan a particular measuring point at periodic intervals. The latter operating method is necessary for calibration but can also be selected to follow up the course or the fluctuations of a given measured value. When this is being done, however, no other measuring cycle can be run.

If it is desired to supplement the average values obtained by normal operation by individual values, and in particular maximum and minimum values, the sensors on the digital recorder can be connected in parallel to an analog recorder and traced by a potentiometer recorder. This simultaneous analog record has the advantage of providing a check on the correct operation of the various sensors.

The data recorded on the punched tape are processed in the Computer Center of the Technical University in Hannover. Fifteen copies are prepared of a daily table showing the most important climatic data from each station. As from 1st October, 1969, the heat balance components were added in the form of a code for each hour, calculated from the stored data (cf. section VI).

# IV. Type of Sensors

The recorder system briefly described above is designed to measure input voltages in the range 0—50 mV; to make full use of this range, the sensors selected for meteorological work have to have an adequate voltage output. This is one of several reasons for not using thermocouples for temperature measurements; this is, in fact, done by means of Pt-resistance thermometers (100 $\Omega$ at 0° C) of specially narrow tolerance. The thermometers used for soil and tree stems (hardened glass resistance thermometers with a sensitive length of 15 mm and a 4 mm diameter, manufactured by Heraeus, Hanau) are unprotected and in direct contact with the medium to be measured. In the near-surface woodland layer of litter and humus it would be easy to record unrepresentative temperatures due to moving sunflecks, so temperature at this level is measured by means of 50 $\mu$ $\varnothing$* Pt wires, encased in a double layer of cellulose triester film and polycarbonate film and laid out over an area of 0.5 m². However, these measuring strips are too sensitive to mechanical damage and as from winter 1969/70 they are replaced by light metal bars 0.5 m long, each containing a resistance thermometer, inserted at various places in the litter/humus layer. Four such bars give a representative temperature for the level in question.

Air temperature and humidity are measured during the season when there are no frosts by means of electrically ventilated psychrometers to Frankenberger's design (Th. Friedrichs, Hamburg). To obtain humidity measurements throughout the year, we used lithium-chloride sensors in winter. This type of sensor has been described by MIESS (1968), although we now use an improved form with a vertical electrode axis.

The resistance thermometers are supplied with constant current; the feed current can be adjusted for each thermometer separately, within narrow limits, to allow calibration to be carried out. Zero-point displacements and angular deviations from the calibration lines can be corrected by feeding in the appropriate summands and factors from the computer. Resolution is 0.05° C. Short-wave radiation (0.3—3 $\mu$) is measured by solarimeters (Kipp & Zonen, Delft). It is not necessary to describe these in detail, as they are standard instruments. The measurement of net radiation, or separate measurements of radiation originating in the upper and lower hemispheres, is done over all wavelengths by means of the Schulze balance meter (B. Lange, Berlin). The instruments used in the Solling have semiconductor elements as receivers and are protected by a dome-shaped lupolene hood. The temperature of the instrument is measured by a resistance thermometer attached to the housing. This instrument, too, need not be described in detail (see FLEISCHER and GRÄFE, 1955/56, and SCHIELDRUP-PAULSEN, 1967).

As the computer of the recorder system can make sense only of positive values received from the digital voltmeter, a 5 mV booster voltage is fed to the radiation input and passed through the amplifier (gain of 20) to the digital voltmeter with 100 mV. This 100 mV is subsequently deducted by the computer to give the actual measured value. This allows negative voltages to be recorded and, by supplying a high input voltage, reduces the digital voltmeter error in the range of lower voltages.

In addition to voltage inputs, the system also accepts impulse inputs, these being generated by the anemometers with inductive pickup used in the Solling. These

---

* $\varnothing$ means diameter.

anemometers have a metal segment mounted on the lower end of the axis of the cup system; this actuates the inductive pickup, producing one impulse per revolution. The impulses are counted and stored, one impulse being equivalent to 2 m run of wind; thus the run of wind is obtained from the number of impulses per unit of time, and when this is divided by (the measuring period in minutes × 30) we obtain the average wind speed for the measuring period. This arrangement largely avoids mechanical losses in the bearing of the cup system and its response sensitivity is 0.2 m/sec.

## V. Disposition of the Sensors

The following values are measured at station W 1:

*Air temperature* at:  10 m; 2 m; 0.5 m (with psychrometers) and at 0.25 m and 0.05 m above ground (with shaded but unventilated thermometers),

*soil temperature* at:  0.01 m; 0.05 m; 0.1 m; 0.2 m; 0.5 m; 1 m below ground,

*air humidity* at:  10 m; 2 m; 0.5 m (with psychrometers and lithium chloride sensors) and 0.05 m (with lithium chloride sensor only) above ground,

*wind speed* at:  10 m; 2 m; and 0.5 m above ground,

*global radiation, sky radiation, albedo, total radiation* from upper hemisphere and total radiation from lower hemisphere at 2 m above ground.

A 10-m high open-frame tower and two 2-m high masts are available for mounting the sensors, which are connected by separate leads to junction boxes on the edge of the measuring field; from here six 40 × 0.8 mm ⌀ cables and a power supply line (220 V transformed on the field to 24 V), buried at a depth of 60 cm, run to the measuring hut 70 m away.

For the sensors at station B 1 a 29.2 m high open-frame tower was erected on a 2 × 2 m base. It is free-standing on concrete foundations and is braced by 4 double steel-wire stays. The tower is equipped with platforms at 4 m intervals, the top one carrying a 7-m high mast on which an instrument-carrying arm can be hoisted by means of a pulley to a height of 35 m. This is the height necessary to record relatively steep gradients in temperature, humidity and wind speed. Despite the uneven density of the stand, the site on which the tower stands gives approximately average conditions and thus the daily sums of, for example, global radiation or net radiation in the stand and on the ground, yield fairly representative values, although the daily figures may be somewhat high due to the presence of moving sunflecks caused by the combination of the sun's altitude and the azimuth.

A boom extending 4.5 m to the south is mounted at a height of 28.6 m to measure radiation conditions above the canopy. The boom can be hauled in over rollers for instrument maintenance.

The following values are measured at station B 1 (see Fig. 1):

*Air temperature* and *humidity* at: 35 m; 28 m; 24.5 m; 20 m; 10 m; 2 m; 0.2 m above (with psychrometers, humidity in parallel with LiCl sensors),

*soil temperature* at: litter humus layer; —0.02 m; —0.05 m; —0.1 m; —0.2 m; —0.5 m; —1 m,

*trunk temperature:* 0—40 mm at 1 m height to east and west; 0—15 mm at 24 m in a branch of 40 mm ⌀,

*wind speed* at: 35 m; 28 m; 26 m; 24.5 m; 20 m; 10 m; 2 m; 0.2 m,

*global radiation* and *total radiation* from the upper and the lower hemispheres at 28.5 m; 24.5 m; 1.8 m; and one movable measuring point,
   *Albedo* at: 28.5 m and 1.8 m.

The program was slightly modified in the summer of 1969 (cf. section VII).

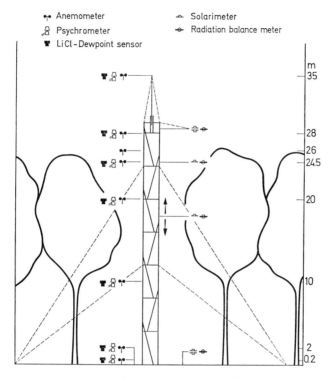

Fig. 1. Arrangement at station B1

# VI. Methods of Determining the Components of the Heat Balance

The data collected are used to calculate the diurnal and annual variations in the components of the heat balance. For determining evapotranspiration two methods are used, thus providing a check on each other. The results of the water balance study supply reference values for longer periods of time.

## 1. The Sverdrup (Bowen) Method

Those components of the heat balance equation

$$Q = -(B + L + V)$$

where $Q$ = net radiation
$B$ = downward heat flow from the soil surface
$L$ = flow of sensible heat
$V$ = flow of latent heat (evapotranspiration)

which cannot be measured directly ($L$ and $V$) are calculated according to SVERDRUP (1935), from the vapour pressure and temperature gradients between two levels above the "outer active surface" on the premises that the diffusivity coefficients of latent and sensible heat are equal.

Most important here is the correct determination of net radiation. Relatively large temperature and vapour pressure gradients will mask the effects of measuring errors and simplifications made during calculation. But generally the temperature and vapour pressure gradients are very small within and above the beech stand. Thus, every effort must be made to have the lowest measuring level as close as possible to the "outer active surface".

We followed Baumgartner's (1956) example in including under one value, $B_P$, the heat flow in the air of the stand $(P_L)$, in the mass of the stand $(P)$ and in the ground $(B)$. The heat balance equation then becomes:

$$Q = -(P_L + P + B + L + V) = -(B_P + L + V).$$

The Bowen ratio $\frac{L}{V}$ is calculated automatically according to an existing program. To the automatic calculation of the heat flow in the soil we are studying whether the relation found by Tajchmann (1967)

$$B_P = -c_1 Q + c_2$$

where $c_1$ and $c_2$ are empirically established constants, is applicable to beech forests. For spruce forest in the Munich area, these constants were:

$$c_1 = 0.125 \text{ and } c_2 = 30 \text{ mcal cm}^{-2} \text{ min}^{-1}.$$

## 2. Determination of Evapotranspiration by Brogmus' Formula

The determination of evapotranspiration being beset with uncertainties, $V$ is calculated in addition by the formula of Monin and Obuchov (1958), as adapted by Brogmus (1959) for any given stratification:

$$V \text{ (mm/h)} = 1.251 \times \frac{f^2 (e_1 - e_2)(u_1 - u_2)}{T_v \left(\ln \left(\frac{z_2}{z_1}\right)\right)^2}$$

$$\text{with } f = 1 - \beta \frac{g}{T_0} \frac{T_2 - T_1}{(u_2 - u_1)^2} \cdot (z_2 - z_1)$$

$g$ = acceleration due to gravity
$T$ = absolute temperature (° K)
$T_v$ = virtual temperature (° K)
$e$ = vapour pressure (mbar)
$u$ = wind speed (cm sec$^{-1}$)
$z$ = height of measurement (cm)
$\beta$ = universal constant (0.6).

For extended periods of time, such a vast amount of calculation is required that it can only be done by a computer. The above formula enables a computer program to be written, as the introduction of the term $f$ eliminates the influence of variations in stratification.

## VII. Preliminary Results

The theoretical concept of energy conversion within a forest demands two conversion levels for sample area B 1 with its distinct vertical structure. The "outer active surface" will be the canopy, the forest floor being an additional "active surface". The radiation conditions illustrated in Fig. 2 (here for short-wave radiation) will show the much greater importance of the crown space as an energy conversion level as opposed to the forest floor. In summer 80.2% of the short-wave radiation is absorbed in the stem and crown space and only 6.8 by the forest floor. The permeability of the stand is naturally greater in winter when the trees are leafless, but even then nearly 60% of radiation is absorbed by trunks and branches and only

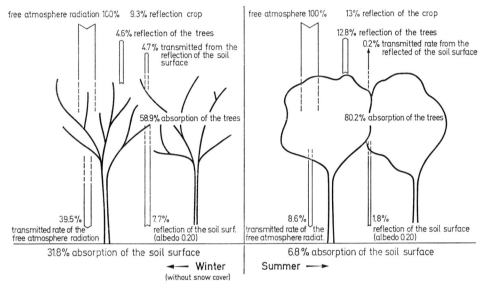

free atmosphere radiation 100%    9.3% reflection crop

4.6% reflection of the trees

4.7% transmitted from the reflection of the soil surface

58.9% absorption of the trees

39.5%
transmitted rate of the
free atmosphere radiation

7.7%
reflection of the soil surf.
(albedo 0.20)

31.8% absorption of the soil surface

⟵ Winter
(without snow cover)

free atmosphere 100%    13% reflection of the crop

12.8% reflection of the trees

0.2% transmitted rate from the reflected of the soil surface

80.2% absorption of the trees

8.6%
transmitted rate of the
free atmosphere radiat.

1.8%
reflection of the soil surface
(albedo 0.20)

6.8% absorption of the soil surface

Summer ⟶

Fig. 2. Short wave radiation in and above the forest (B1)

31.8% by the forest floor. Above fresh snow, which has an albedo of 0.90, the proportion of radiation absorbed by the timber mass of the stand will rise almost to the summer value (77.5%). This shows that short-wave radiation is absorbed not only on the way into the stand but also on the way out (by reflection), and so contributes to the total radiation adsorbed by the timber mass, even leaving out of account the multiple reflection within the stand.

Table 2. *Amount of short-wave radiation absorbed in the beech stand*

| height above forest floor (m) | % of total absorption within the beech stand B1 |
|---|---|
| 28.5—27 | 7.5 |
| 27  —25 | 20.0 |
| 25  —24.5 | 14.0 |
| 24.5—23 | 25.5 |
| 23  —21 | 10.0 |
| 21  —17 | 10.0 |
| 17  — > 0 | 6.2 |
| at the forest floor | 6.8 |

The concentration of the foliage mass in summer at a height of 20 to 26 m led us to expect that the canopy would not act homogeneously as the "outer active layer", but that this would be concentrated in its upper region. Therefore, during the summer of 1969 we tried to determine the exact location of this active layer by placing sensors relatively close together at heights of 29 m, 27 m, 25 m and 22 m.

Particular care was taken to measure the amount of radiation penetrating into the canopy at various depths. The result is shown in Table 2.

The permeability to short-wave radiation is modified by the relative proportions of solar and sky radiation in the global radiation. At a height of 24.5 m on sunny days (global radiation $>600$ ly d$^{-1}$) only 50%, but on overcast days (global radiation $<200$ ly d$^{-1}$) more than 70% of the sum of the radiation measured above the stand will be absorbed. Therefore, although the amounts shown in Table 2 are only approximated, they do show very clearly that the most active region of radiation conversion lies between 24 and 25 m.

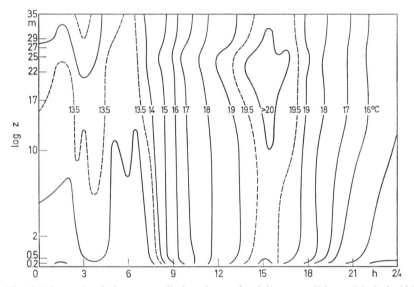

Fig. 3. Thermoisopleths on a radiation day under foliage conditions (B1, 9. 9. 1969)

The importance of the canopy layer between 24 and 25 m, as identified by radiation measurements, is confirmed by the diurnal variations in temperature within and above the stand on a sunny day (Fig. 3). During the day the effect of the main conversion layer is apparent in excess temperatures at this level. At night, however, a nocturnal type of radiation is established throughout the stand as a result of the subsidence of cold air from the canopy, but these phenomenona are very much less than the well-known profiles found in open country.

The occurrence of very slight gradients of air temperature in and above the stand during the day must be emphasized. Heckert (1959) also found this phenomenon in an oak forest near Potsdam and this led him to reject the concept of an "interface" between the stand and the free atmosphere, in spite of the undoubted existence of an "active surface".

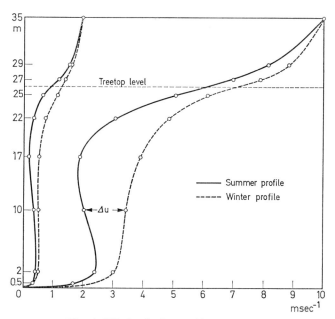

Fig. 4. Wind velocity profile (mean, B1)

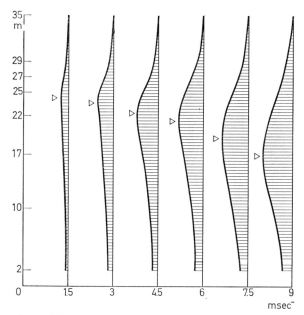

Fig. 5. Summer-winter difference of wind velocity ($= \Delta u$) for distinct degrees of velocity at the 35 m level in B1 ($\triangleright = \Delta u$ max)

Complementary to that, I ascribe this phenomenon particularly to the simultaneous heating of the trunks, branches and leaves by radiation absorption at all heights of the stand, but preferentially the upper canopy region.

This statement is based on the fact that the measurement of wind speed in vertical profile shows a clear decline as the wind penetrates into the stand. There is a "winter profile" as distinct from a "summer profile". With full foliage the slackening of horizontal wind speed within the stand is greater than in winter. In summer the wind speed rises again below the canopy and attains a second maximum at 2 to 5 m height. During winter, however, wind speed falls constantly — but not evenly — with depth of penetration (Fig. 4). Comparing winter and summer wind-speed profiles, we observe that the greatest difference in speed moves to lower levels with increasing speed. At a wind speed of 1.5 m sec$^{-1}$ (measured at 35 m) the greatest difference occurs at 25 m, with 9 m sec$^{-1}$, at about 17 m (Fig. 5).

The results described here are of a preliminary nature, and they should be considered as preparatory investigations for the computation of the heat balance. A detailed discussion would therefore be premature.

# References

Baumgartner, A.: Untersuchungen über den Wärme- und Wasserhaushalt eines Waldes. Ber. deut. Wetterdienst 5, No. 28 (1956).
— Wärmeumsätze des Bodens und der Pflanze. In: F. Schnelle (Ed.): Frostschutz im Pflanzenbau, Vol. 1. München-Basel-Wien: BLV-Verlagsgesellschaft 1963.
Brogmus, W.: Zur Theorie der Verdunstung der natürlichen Erdoberfläche. Deut. Wetterdienst/Seewetteramt. No. 21, Hamburg 1959.
Fleischer, R., Gräfe, K.: Die Ultrarot-Strahlungsströme aus Registrierungen des Strahlungsbilanzmessers nach Schulze. Ann. Meteorol. 7, (H. 1/2), (1955/56).
Galoux, A.: Flux d'énergie et cycles de matières en tant que processus écologiques. Extr. Ass. Nation. Professeurs Biol. Belgique. 14e Année, 1968, No. 4.
Heckert, L.: Die klimatischen Verhältnisse in Laubwäldern. Z. f. Met. Berlin 13 (1959).
Miess, M.: Meßfehler bei der Taupunktmessung mit Lithiumchlorid-Feuchtefühlern. Arch. Met. Geophys. Bioklimatol. Ser. B, 16 (1968).
Monin, A. S., Obuchow, A. M.: Fundamentale Gesetzmäßigkeiten der turbulenten Durchmischung in der bodennahen Schicht der Atmosphäre. In: Goering, H. (Ed.): Sammelband zur statistischen Theorie der Turbulenz. Berlin: Akademie-Verlag 1958.
Schieldrup-Paulsen, U.: Some experiences with the calibration of radiation balance meters. Arch. Met. Geophys. Bioklimatol. Ser. B, 15 (1967).
Sverdrup, H. U.: The eddy conductivity of the air over a smooth snow field. Geophys. Publ. 1, No. 7 (1935).
Tajchmann, S.: Energie- und Wasserhaushalt verschiedener Pflanzenbestände bei München. Wiss. Mitt. Met. Inst. Univ. München 12 (1967).

# T. The Characterization of the Woodland Light Climate

W. Eber

## I. Introduction

The measurement of light has posed very difficult problems of environmental research on the woodland floor because of its continuous variation in intensity, spatial variation and spectral composition (Evans, 1956). Many attempts have been made to find a suitable way of characterizing the woodland light climate. The most common has been to express light measurements as a percentage of light in the open. Unfortunately this principle has been applied in different ways.

Relative values are usually obtained by instantaneous measurements under cloudy conditions. But measurements made under sunny conditions, too, are expressed as the ratio of diffuse light inside the wood to either diffuse light (Atkins and Poole, 1926; Filzer, 1939) or total light (Wagner, 1938; Tchou, 1948) in the open. Even values integrated over clear days (Logan, 1966) or over longer periods with different weather conditions (Röhrig, 1967) have been used to calculate relative values.

It has been pointed out repeatedly that relative values differ with weather conditions (Walter, 1933; Evans, 1956; Schlüter, 1966) and the spectral properties of the receiver (Atkins, Poole and Stanbury, 1937). Furthermore, instantaneous measurements exclude direct sunlight and do not take account of the seasonal variation of the light climate (Anderson, 1964). They also give no clue to the absolute light intensities on which photosynthesis depends (Bormann, 1956).

On the other hand, distinct correlations between the spatial distribution of plants and definite ranges of intensity have only been demonstrated for diffuse light (Wiesner, 1907; Trapp, 1938; Ellenberg, 1939; Blackman and Rutter, 1946). Thus it is worthwhile to give further attention to this method, especially as it is very simple and less costly and troublesome than integrating measurements with recording instruments. Therefore the aim of this study was to investigate the dependence of percentage values and spatial distribution of the diffuse light on weather type and solar altitude and to compare them with results obtained by integrating measurements.

## II. Light Measurements

### 1. Integrating Measurements

The integrating measurements were made in the year 1968 in the beech sample plot B 1 using selenium photocells (Dr. B. Lange) in connection with a three colour track chopper-bar recorder. One cell (A) was mounted on the instrument tower above the canopy, two cells (B and C) 50 cm above the ground. The positions of the cells were: cell B under a canopy hole above a group of *Luzula albida*, cell C under a denser part of the canopy above bare ground. During the highest solar altitudes cell C received much and cell B only a little direct sunlight. It should be mentioned that the sensitivity of the cells, which had already been used the year before, decreased by less than 5% during the whole growing season. The daily totals were obtained by planimetry of the recordings.

## 2. Instantaneous Measurements

Percentage values for cloudy days were calculated from the tracks of the recorder. On clear days the measurements were made with a Dr. B. Lange standard photometer at the site, and immediately before and after on the instrument tower above the canopy. Where only diffuse light was wanted, the direct sunlight was shaded off by a small paper disc. This might cause errors when near noon canopy holes important for the penetration of diffuse light are obstructed.

## 3. Mapping the Spatial Distribution of Diffuse Light

The distribution of the diffuse light, expressed as relative light intensity, was mapped on both overcast and clear days from dawn to dusk, following the method of ELLENBERG (1939).

Eleven parallel lines were laid at 1 m spacing across a 10 by 10 m sampling plot and measurements were made along these lines successively at 1 m intervals. On overcast days each single value could be related to that recorded at the same time from above the canopy, whereas on clear days the values of the single measurements had to be corrected for the variation of the diffuse light intensity continuously observed at one point within the plot. To facilitate comparison of the different maps, the values of this point of comparison were set equal, the value for overcast sky, being taken in every case.

## III. Some Results

### 1. Instantaneous Measurements

The percentage values of uniform overcast sky proved rather constant, independent of the time of the day. Values of days with irregular cloud distribution must be treated with caution, because with greater irregularities they differ considerably from those of overcast days (Table 1), though the variance of a series of observations may be rather small. But averaged over longer periods, they are similar to those of uniform overcast sky. Therefore percentage values calculated from totals of cloudy days also differ only slightly from values for overcast sky (Fig. 2a).

On clear days a characteristic curve of the relative light intensity can be observed (Fig. 1b). Its form is determined by the fact that the brightest region of diffuse light in the sky is around the sun and that, as the sun's altitude increases, an increasing proportion of light from this region penetrates the canopy, the density of which is known to decrease from horizon to zenith. The rapid increase near noon of the ratio of diffuse light in the wood to diffuse light in the open is probably due to a high proportion of light reflected by or transmitted through leaves. This light is rich in those wavelengths in which the photocells have their maximum sensitivity. In the morning the values are lower, at noon much higher than those for overcast days.

Table 1. *Relative light intensities during different periods measured at 3-minute intervals in the beech stand B1*

(number of observations)

| relative light intensity % | irregularly cloudy | | | | | | | | | | uniformly overcast |
|---|---|---|---|---|---|---|---|---|---|---|---|
| time of the measurement (1967) | 29.7. 6.30–8.00 | 28.7. 14.15–16.15 | 5.7. 10.10–11.20 | 20.6. 16.45–18.50 | 8.7. 15.00–17.20 | 20.6. 8.25–10.40 | 3.7. 8.30–12.00 | 14.7. 6.50–10.00 | 24.8. 14.15–16.15 | total | 22.8. 9.00–18.30 |
| 3.5 | | | | | | | | | 1 | 1 | |
| 3.6 | | | | | | | 1 | | 5 | 6 | |
| 3.7 | | | | | | | | 1 | | 1 | |
| 3.8 | | | | 1 | | | 2 | 5 | 7 | 15 | |
| 3.9 | | | | 1 | | | 2 | 3 | 5 | 11 | |
| 4.0 | | | | | | | 3 | 2 | 9 | 14 | |
| 4.1 | | 1 | | | | | 3 | 3 | 4 | 11 | 1 |
| 4.2 | | | | | 1 | 2 | 3 | 5 | 2 | 13 | |
| 4.3 | | | | 3 | 1 | 1 | 3 | 6 | 1 | 15 | 6 |
| 4.4 | | 1 | | 3 | 3 | 4 | 4 | 1 | | 16 | 2 |
| 4.5 | | 1 | 1 | 1 | 2 | 4 | 7 | 2 | | 18 | 2 |
| 4.6 | | 1 | 2 | 1 | 5 | 4 | 4 | 5 | | 22 | 7 |
| 4.7 | | 2 | | 1 | 7 | 6 | 7 | 9 | | 32 | 11 |
| 4.8 | | 1 | 2 | 7 | 4 | 3 | 3 | 5 | | 25 | 26 |
| 4.9 | | 1 | 1 | 4 | 3 | 4 | 6 | 7 | | 26 | 26 |
| 5.0 | | 2 | 3 | 4 | 6 | 1 | 6 | 2 | | 24 | 57 |
| 5.1 | | 3 | 1 | 1 | 5 | 6 | 3 | | | 19 | 28 |
| 5.2 | 1 | 3 | 1 | 1 | 5 | 8 | 2 | 2 | | 23 | 11 |
| 5.3 | 3 | 5 | | 2 | 4 | 3 | 2 | 1 | | 20 | 9 |
| 5.4 | 1 | 1 | 3 | 3 | 2 | 1 | 3 | | | 14 | 2 |
| 5.5 | 3 | 3 | 1 | 1 | | | 4 | | | 12 | 1 |
| 5.6 | 2 | 1 | 2 | 2 | | | | | | 7 | |
| 5.7 | 2 | 1 | 1 | 1 | | | 2 | | | 7 | |
| 5.8 | 3 | 2 | | 1 | | | | | | 6 | |
| 5.9 | 1 | 1 | 1 | 1 | 2 | | 1 | | | 7 | |
| 6.0 | 2 | 2 | 1 | 1 | | | 1 | | | 7 | |
| 6.1 | 6 | | 1 | | | | | | | 7 | |
| 6.2 | 3 | | 2 | 2 | | | | | | 7 | |
| 6.3 | 1 | 2 | 1 | 2 | | | | | | 6 | |
| 6.4 | 1 | 4 | | | | | | | | 5 | |
| 6.5 | | 2 | | | | | | | | 2 | |
| 6.6 | | 1 | | | | | | | | 1 | |
| 6.7 | | 1 | | | | | | | | 1 | |
| 6.8 | 1 | | | | | | | | | 1 | |
| mean value (%) | 5.85 | 5.51 | 5.37 | 5.12 | 4.91 | 4.85 | 4.73 | 4.47 | 3.92 | 4.90 | 4.94 |
| standard deviation | 0.38 | 0.67 | 0.55 | 0.61 | 0.48 | 0.47 | 0.53 | 0.40 | 0.12 | 0.69 | 0.23 |

The same absolute values related to total light in the open result in far lower relative values. In the morning these values decrease rapidly when direct sunlight hits the cell in the open. Towards noon reflected and transmitted light increase so enormously that even the percentage values increase slightly until noon.

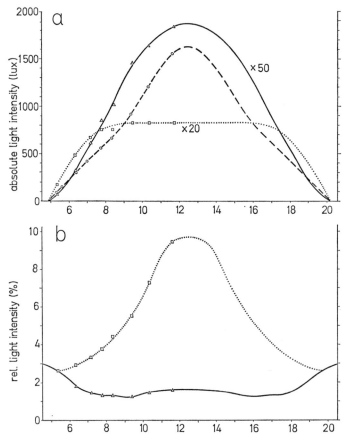

Fig. 1. Daily course of absolute and relative light intensity on 30. July. a) Daily course of total light in the open (solid line), diffuse light in the open (dotted line) and in the wood (interrupted line). b) Daily course of diffuse light in the forest as percentage of diffuse (dotted line) and total light (solid line) in the open. The curves completed for the afternoon from values of other days are in reality only approximately symmetrical

With decreasing seasonal altitude of the sun these curves become lower. Their height is also influenced by the ratio of diffuse light in the open to total light in the open. As this proportion increases the relative values related to diffuse light in the open decrease, whereas those related to total light in the open increase.

## 2. Integrating Measurements

Percentage values calculated from daily totals are similarly dependent on weather conditions (Fig. 2a and b). As already mentioned, measurements on cloudy days

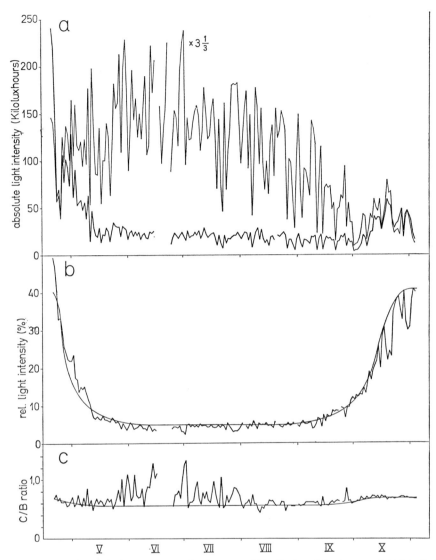

Fig. 2. a) Daily totals for the season in the open (thin line) and in the wood (thick line) at the position of cell B. b) Relative light intensity calculated from the totals in Fig. 2a. The smooth curve shows the values for uniformly overcast sky. c) Ratio of the light intensity recorded by cell B (little direct sunlight) to that of cell C (much direct sunlight). The smooth curve shows the ratio for uniformly overcast sky

gave values similar to those obtained by instantaneous measurements under uniform overcast conditions. The relative values for clear days decrease with the seasonal altitude of the sun. From May to July they depend upon the quantity of direct sunlight received at the particular site, being slightly or even much higher than those of cloudy days, while at the end of the growing season they are far lower (Fig. 2c, Table 2).

10*

Relative values derived from the totals for the whole growing period may be higher
or lower than values for overcast days, according to number and temporal distribution of sunny days, and therefore differ from year to year.

The seasonal course of the absolute light intensity (Fig. 2a) is entirely different
from that of the relative values (Fig. 2b). When the trees come into leaf, the absolute

Table 2. *Results of integrating measurements in 1968*

| | A | | B | | | C | | |
|---|---|---|---|---|---|---|---|---|
| | a | b | a | b | c | a | b | c |
| 21.—30. April | 3474 | 5.2 | 1162 | 20.2 | 33.3 | 958 | 19.3 | 27.5 |
| May | 13408 | 20.1 | 1284 | 21.8 | 9.6 | 1040 | 21.0 | 7.8 |
| June | 15275 | 23.0 | 641 | 11.2 | 4.2 | 701 | 14.2 | 4.6 |
| July | 12467 | 18.7 | 621 | 10.8 | 5.0 | 597 | 12.0 | 4.8 |
| August | 10779 | 16.1 | 563 | 9.8 | 5.2 | 404 | 8.2 | 3.7 |
| September | 7186 | 10.7 | 457 | 8.0 | 6.4 | 354 | 7.1 | 4.9 |
| October | 3842 | 5.7 | 924 | 16.1 | 24.1 | 798 | 16.2 | 20.8 |
| 1.—4. November | 330 | 0.5 | 119 | 2.1 | 36.0 | 100 | 2.0 | 30.4 |
| 21. April—4. Nov. | 66761 | 100.0 | 5771 | 100.0 | 8.6 | 4952 | 100.0 | 7.4 |

The position of the cells A, B and C see in the text. a) Light totals in kilolux-hours;
b) Proportion of the light totals of the whole season; c) Relative intensity calculated from
the light totals of the given period.

light intensity inside the wood does not decrease very rapidly, because the light totals
in the open are still increasing. Whereas during summer the relative values are rather
constant, the absolute values show considerable variation and decrease continuously
with the seasonal decrease of the sun's altitude. When the leaves fall, the relative
values increase much more than the absolute values, whose increase is limited by the
gradual decrease of light totals in the open. In general, direct sunlight tends to even
out illumination differences in the distribution of diffuse light on the woodland floor.
Whereas under cloudy conditions the brightest places are found below canopy holes,
the sunflecks occur according to the sun's altitude in places which are usually darker.

### 3. Maps of the Distribution of Diffuse Light

The distribution of diffuse light in the sampling plot (Fig. 3a—k) is very similar
on overcast and clear days at low altitudes of the sun, although in the first case the
differences in intensity within the plot are greater. On clear days the distribution of
light changes towards noon: the brightest regions move to the north and the areas
of equal brightness tend to extend in the direction of the sunlight, while at the same
time the differences in brightness become smaller.

## IV. Discussion and Conclusions

The usefulness of any method is judged by whether the results are reproducible; in this case, it must also allow different sites to be compared and explain their meaning for the vegetation. The percentage values for clear or irregularly cloudy conditions obviously do not meet these demands. The relative values for uniformly overcast sky, however, proved reproducible and suitable for the characterization of sites and their comparison with others. Their numerical value does not represent the absolute quantity of light, but is only an approximate reflection of the canopy density at a given time. The good correlation between percentage values of overcast periods that has been observed by several authors may be attributed to the fact that in our climate the duration of sunflecks is severely limited by cloudiness (ATKINS, POOLE and STANBURY, 1937). Moreover, their effect on photosynthesis is weakened by the low light saturation of woodland plants and can even be negative when the transpiration is raised too strongly (ELLENBERG, 1963, p. 216).

It is also important that before the leaves are fully developed, a period which contributes a high proportion to the totals of the growing season, the light distribution on overcast days is rather similar to that in the summer (Fig. 3a) and that the spatial differences of both diffuse and total light are much smaller than in summer (Fig. 2c).

Maps of the distribution of diffuse light combined with percentage values for uniform overcast periods may give useful information about the spatial variation of diffuse light. It is regarded as a great advantage that maps of this kind, previously made only under cloudy conditions (TRAPP, 1938; ELLENBERG, 1939; BLACKMANN and RUTTER, 1946), can also be made on clear days, when the diffuse light intensity changes so slowly that measurements can readily be carried out by one person without recording instruments. The simplicity and low cost of this method make it especially suitable for minimum programs.

Nevertheless, investigations taking in all aspects of the light climate cannot neglect the contribution of direct sunlight and therefore need recording instruments. But it is difficult to interpret the meaning of daily or seasonal totals for the growth of plants, as plants respond to particular intensities over specific periods of time. Therefore, according to BORMANN (1956), complete information on this subject can only be acquired by analysing aspects of intensity duration. This will be the subject of a further study.

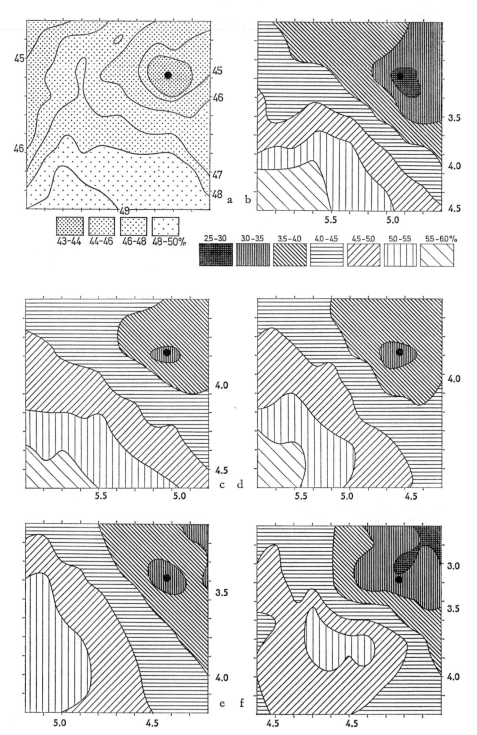

Fig. 3 a—f

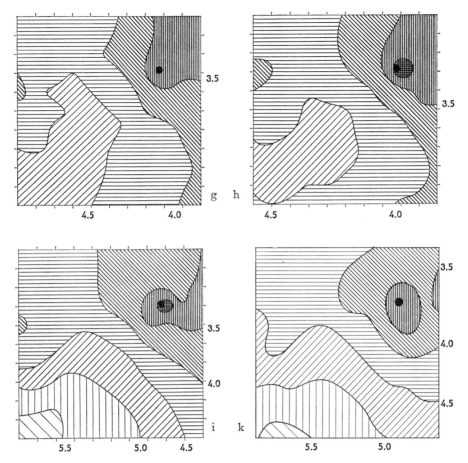

Fig. 3a—k. Maps of the distribution of diffuse light on uniformly overcast days in winter (a) and summer (b) and on clear days in summer (c—k). a) 18. 11. 1968 11.30 h; b) 20. 8. 1968 6.45 h, c) 31. 7. 1968 8.00 h; d) 31. 7. 1968 10.00 h; e) 30. 7. 1968 11.00 h; f) 1. 7. 1968 12.15 h; g) 30. 7. 1968 12.45 h; h) 28. 8. 1968 14.00 h; i) 28. 8. 1968 18.00 h; k) 19.8.1968 19.20 h [each 10 × 10 m]

# References

ANDERSON, M. C.: Studies of the woodland light climate. II. Seasonal variation in the light climate. J. Ecol. 52, 643—663 (1964).

ATKINS, W. R. G., POOLE, H. H.: Photoelectric measurements of illumination in relation to plant distribution. Part. I. Sci. Proc. Roy. Dublin Soc. 18, 277—298 (1926).

— — STANBURY, F. A.: The measurement and the colour of the light in woods by means of emission and rectified photoelectric cells. Proc. roy. Soc. B., 121, 427—450 (1937).

BLACKMAN, G. E., RUTTER, A. J.: Physiological and ecological studies in the analysis of plant environment. I. The light factor and the distribution of the bluebell *(Scilla non-scripta)* in woodland communities. Ann. Botany N. S. 10, 361—390 (1946).

BORMANN, F. H.: Percentage light readings, their intensity-duration aspects, and their significance in estimating photosynthesis. Ecology 37, 473—476 (1956).

Ellenberg, H.: Über Zusammensetzung, Standort und Stoffproduktion bodenfeuchter Eichen- und Buchen-Mischwaldgesellschaften Nordwestdeutschlands. Mitt. Florist.-Soziol. Arb. gem. Niedersachsen **5**, 3—135 (1939).
— Vegetation Mitteleuropas mit den Alpen. Stuttgart: E. Ulmer 1963.
Evans, G. C.: An areas survey method of investigating the distribution of light intensity in woodlands, with particular reference to sunflecks. J. Ecol. **44**, 391—428 (1956).
Filzer, P.: Lichtökologische Untersuchungen an Rasengesellschaften. Botan. Cbl. **60**, 229—248 (1939).
Logan, K. T.: Growth of tree seedlings as affected by light intensity. II. Red Pine, White Pine, Jack Pine and Eastern Larch. Dep. For., Publ. No. 1160, 1966.
Röhrig, E.: Wachstum junger Laubholzpflanzen bei unterschiedlichen Lichtverhältnissen. Allgem. Forst- u. Jagdztg. **138**, 224—239 (1967).
Schlüter, H.: Licht- und Temperaturmessungen an den Vegetationszonen einer Lichtung („Lochhieb") im Fichtenforst. Flora Abt. B, **156**, 133—154 (1966).
Tchou, Y. T.: Études écologiques et phytosociologiques sur les forêts riveraines du Bas-Languedoc. II. Vegetatio **1**, 93—128 (1948—49).
Trapp, E.: Untersuchungen über die Verteilung der Helligkeit in einem Buchenbestand. Bioklim. Beibl. Meteorol. Z. **5**, 153—158 (1938).
Wagner, H.: Influence de la lumière sur la repartition locale des espèces de quelques associations méditerranéennes. Comm. Station Internat. Géobot. Médit. et Alp. **51**, 1938.
Walter, H.: Ein neuer für botanische Exkursionen geeigneter Lichtmesser. Ber. dtsch. Botan. Ges. **62**, 362—376 (1933).
Wiesner, J.: Der Lichtgenuß der Pflanzen. Leipzig: Lehmann 1907.

# U. Aspects of Soil Water Behavior as Related to Beech and Spruce Stands
## Some Results of the Water Balance Investigations

P. Benecke and R. Mayer

## I. Introduction

One of the problems to be investigated in the Solling Project is the water balance as related to spruce and beech stands within the same environment. The term environment here includes both the climatic and soil conditions, described briefly below.

Knowledge of the water balance provides information regarding: 1. the rate of transpiration, which is closely related to plant production, and 2. the rate of vertical flow beneath the root zone, thought to be identical with the rate of ground water recharge.

The basic considerations as to how to measure it are given by the water balance equation

$$P = R + S + I + ET + \Delta St$$

where  $P$ = precipitation
$R$ = runoff
$S$ = seepage
$I$ = interception
$ET$ = evapotranspiration
$St$ = storage

Interest will be focussed in this paper on investigating the behavior of the water in the soil, and this means essentially obtaining sufficient information about the term "$S$" in the above equation. Since the sites are almost level, "$R$", the surface runoff, may be neglected. "$I$" is measured as the difference between precipitation on a nearby open site and the sum of precipitation (canopy drip) plus stem flow per unit area within the forest stand. The problems arising in this connection are not discussed here. "$\Delta St$", the change in storage, is measured by means of tensiometers* using the pF-curve** to convert the data. Furthermore, a neutron moisture meter is used. Eventually the equation may be solved for "$ET$", i.e. this term is derived from all the other members of the equation. Hence any error made in the measurements will appear in $ET$ also. The greatest difficulty arises in the estimation of "$S$". The concept is primarily to measure tension gradients at various depths of the soil as related to the

---

* We thank Prof. F. Richard, Zürich, who was kind enough to demonstrate a site near Zürich used for water balance investigations, thus suggesting how to design the tensiometers used in these investigations. For details see Brülhart, 1969.

Tensiometers are frequently described in the literature. A comprehensive description of principles and design may be found e.g. in Rose (1966).

** That is the ratio of water content to water tension.

soil and subsoil horizons. Since there are many sources causing inhomogeneities (irregular rainfall due to the trees, branches etc., stem flow, withdrawal by irregularly distributed roots and soil inhomogeneities), one has to assess the statistical significance of the data. To do this, a large number of measurements need to be made.

## II. Climate, Geology and Soil

The investigation sites are located on a large plateau, consisting mainly of Triassic sandstone (Buntsandstein), from about 400 to 500 m above sea level. For the Solling area the annual rainfall is about 1100 mm with rather uniform distribution. The average temperature is 6.5 °C. One of the two investigation sites (B 1) is covered by beech of age 120 years and the other by spruce of age 85 years (F 1). The soils have been developed partly in evidently para-autochthonic solifluction masses originating from periglacial processes in the weathering top layer of the "Buntsandstein" and partly in a loess layer covering most of that area. The loess, the main constituent of the soil, is considered to be a recent form possibly taken up from older loess deposits and resedimented.

This layer also is considered to be a para-autochthonic creeping earth. Its striking features are strong acidity and very low bulk density combined with high physical stability. Neither breakdown of the high-pore-space structure nor development of distinct soil horizons occurs, except for a very slight podsolization in the very top layer.

The soil is denoted an acid "Braunerde". The description of the profile reads as follows, starting with the beech site

| horizon* | depth (cm) | description |
|---|---|---|
| L | +3.5 to +1.8 | un decomposed litter, moderately stratified, gradual boundary |
| F | +1.8 to +0.3 | residues of plant tissues mixed with completely decomposed organic matter, the latter increasing with depth, stratification more distinct, diffuse boundary |
| H | +0.3 to 0.0 | completely decomposed organic material containing a certain amount of residues of plant tissues, diffuse smooth boundary |
| A$_h$ | 0.0—1.0 | clayey silt, weakly stony, dark gray, humous, many roots; weak platy structure, few to common pores, slightly plastic, very low density; gradual slightly irregular boundary |
| A$_{eh}$ | 1—8 | clayey silt, weakly stony, pale reddish brown gray (podzolization!), locally some mottling, weakly humous, decreasing; many roots, irregular weak platy structure partly subangular blocky, very low density, diffuse boundary |
| B$_v$ | 8—60 | clayey silt, locally weakly stony, yellowish brown, plentiful roots, weak medium subangular blocky, slightly medium plastic, low density, common pores, clear wavy boundary |

* Horizon symbols according to „Arbeitsgemeinschaft Bodenkunde: Die Bodenkarte 1:25 000." For more detailed description of L, F and H see Babel, chapter Va.

| $B_v/II\ B_v$ | $\pm 60—85$ | silty loam, stony, pale brownish-gray, few fine roots, decreasing; angular blocky (locally weakly platy), common moderately thick clay films, plastic, few pores, medium density, clear wavy boundary |
| II $B_v$ | 85—140 | loam, plentiful stony, increasing; brownish-red, no roots, angular blocky to massive, common moderately thick clay films, weakly cemented, very few pores, very high density, diffuse boundary |
| II/III | below 140 | transition to weathered zone of the "Buntsandstein"; the platy stones are still in the original position the interstices between them firmly filled up with brownish-red loam. Below about 180 cm the loam filling decreases, thus gradually increasing the permeability |

Since there is little difference between the soil profiles under beech and spruce as far as the mineral soil is concerned, only the differences in the organic layer need be pointed out. The thickness of this layer under spruce varies between 3 and 7 cm. Furthermore the stratification is more distinct and the boundary at the bottom of the H layer is far more distinct and may be called "abrupt".

## III. Experimental Layout

In designing the arrangement of the measuring devices, the presumed effects of the trees and the soil horizons have to be taken into account. Fig. 1 shows that the

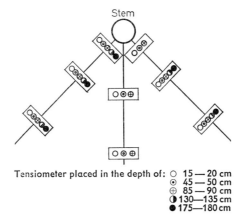

Tensiometer placed in the depth of:  ○ 15 — 20 cm
⊙ 45 — 50 cm
⊕ 85 — 90 cm
◐ 130—135 cm
● 175—180 cm

Fig. 1. Arrangement of the tensiometers

region around a tree representing the average of one group of trees in the stand is divided into annular subregions representing the principal zones of precipitation gain: the subregion immediately surrounding the stem is influenced by stem flow, one or two annular subregions control the water gains under the canopy and a further subregion receives the main part of the canopy drip or throughfall. Each subregion is represented by a section of an annulus.

Table 1 shows in case of spruce the well-known fact (Eschner, 1965) of increasing canopy drip with increasing distance from the stem. This does not hold true for beech due to the substantial effect of stem flow. Since the data are averaged over periods with very different moisture conditions, only a rough tendency should be inferred. The figures given in Table 1 result from least-squares analysis (see below).

Table 1. *Soil-water-tension response to the distance from the stem;*
*(data are averaged over time — June to Oct. 1968 —, depths and plots, in cm water column)*

|        | area of crown projection near to the stem | canopy drip | outside crown projection |
|--------|-------------------------------------------|-------------|--------------------------|
| beech  | —42.3                                     | —58.8       | —43.6                    |
| spruce | —94.2                                     | —64.8       | —47.6                    |

Since water movement is governed by hydraulic gradients and a main component in setting up these gradients is due to matrix potentials, tensiometers are used. As indicated in Fig. 1 they are distributed over 5 depths of the soil in such a way as to measure either the upper and lower tension of a layer regarded as sufficiently homogeneous with respect to permeability (15 to about 50 cm and 125 to about 180 cm), or to measure a medium value (about 80 cm depth). Particular interest must be focussed on the region below the root zone in order to obtain the net downward water flow, assumed to be identical with the ground water recharge, the most problematic term in the water balance equation.

Altogether almost 500 tensiometers have been inserted. Readings are done usually three times a week, the particular timing being correlated with the rainfall distribution. The readings are listed before being transferred onto punched cards; they are then fed into a computer which produces a list of all data and — after computing mean values and some standard statistics — a second list containing mean values as related to the various depths, stands (beech or spruce), plots within the stands, position within the plots (i.e. the tensiometers grouped by the annular subdivision).

# IV. Soil Water Tension

## 1. Changes in Tension with Time

A further table is set up by the computer, arranging the data in a kind of coordinate system, the ordinate representing the depth below surface and the abcissa showing the time. Fig. 2 represents a period. To give a comprehensive view of the distribution the individual tensiometer readings are omitted and only the regions of readings belonging to the same interval of soil water tension are displayed. It may be pointed out that in general soil water content reaches a maximum in spring after the snow has melted and a minimum in autumn. Despite this the period shown in Fig. 2 includes both maximum and minimum soil moisture tension data and thus soil moisture content changes as observed so far in this project. The only exception with higher tension readings is summer and fall 1969, a period which was extremely hot and dry in Central Europe.

Fig. 2 illustrates the very important fact that almost no change occurs in moisture conditions in the root-free loamy subsoil of very high bulk density, despite considerable changes in the upper horizons. Note that the rainfall is indicated at the top of the figure. In addition a general tendency towards a drier soil under spruce can be seen particularly under relatively wet conditions.

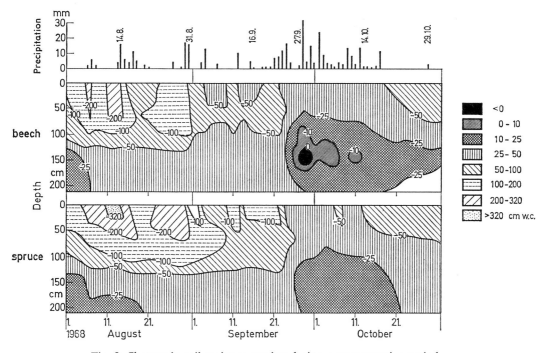

Fig. 2. Changes in soil moisture tension during a representative period

To check this important tendency, the least-squares analysis was used again. The result is shown in Fig. 3 for the time interval June 1st to Nov. 15th, 1968. A considerably higher average soil water tension under spruce is indicated.

## 2. Comment on Statistical Methods Used

Before continuing the discussion about differences of soil water tension gradients under beech and spruce, a brief comment on the statistical methods employed may be useful. As compared with the analysis of variance, the least-squares method (HARVEY, 1960) has the advantage of being independent of uniform class frequencies of the data, the latter being highly desirable when using analysis of variance. The least-squares analysis provides a mean value for the total of all data and the deviations of the mean values of the subclasses. The overall mean value is computed so as to make the sum of squares of the deviations a minimum. The significance of the differences has not been calculated so far, though it is possible, but rather involved. Instead the least-squares analysis has been used in order to obtain time-averaged results to show the general tendencies.

On the other hand, it would be desirable to investigate typical situations, which may serve to represent certain intervals and may eventually allow the computation of flow rates.

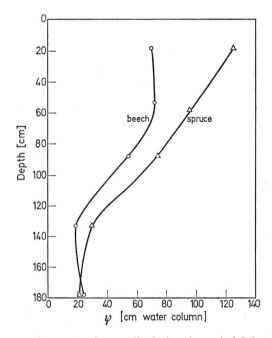

Fig. 3. Average soil water tension profile during the period July to October 1968

For this, the significance of the mean values as related to the tree species (effect A), the plots $[B(A)]$, the position within the plot $[C(B)]$ and the depth $(D)$, and the significance of the differences between these values, should be known. Thus an analysis of variances was carried out. The model underlying the analysis is shown in Table 2. It is a mixed model, i.e. a two-way classification with two hierarchic subgroups. It should be mentioned that in order to employ the analysis of variance the class frequencies have been corrected, i.e. some values have been omitted and other groups have been completed by adding data according to their highest probability.

## 3. Typical Tension Profiles

Table 2 represents some of the results for four selected days with different moisture conditions. Besides the mean values, the components of variance and the results of the significance test can be seen.

Emphasis may be put on the non-existence of significant differences between plots within sites, thus allowing all the data for one depth to be pooled in order to compare the spruce and the beech stands.

Table 2. *Mean values, components of variance, and results of F test on four selected days*

| Date | Soil moisture conditions | Overall mean value in cm w. c. | Main effects A sites | B (A) plots | C (B) positions | D depths | Interactions A × D | B (A) × D | C (B) × D | Error E |
|---|---|---|---|---|---|---|---|---|---|---|
| 28. 9. 1968 | "very moist" | 20.5 | * 6.98 | n. s. 0 | *** 6.53 | *** 11.56 | n. s. 0 | n. s. 0 | * 7.32 | 16.73 |
| 30. 10. 1968 | "moist" | 42.4 | ** 4.07 | n. s. 0 | *** 4.60 | n. s. 16.90 | ** 10.43 | n. s. 0 | ** 5.48 | 8.72 |
| 11. 8. 1968 | "dry" | 129.8 | n. s. 0 | n. s. 0 | *** 43.05 | * 98.24 | *** 27.53 | n. s. 0 | n. s. 37.55 | 104.82 |
| 13. 8. 1968 | "very dry" | 151.5 | n. s. 10.93 | n. s. 0 | *** 59.24 | * 117.29 | *** 21.97 | n. s. 0 | n. s. 34.94 | 109.74 |

It is, of course, most interesting to check differences between the various depths for spruce and beech and to determine their significance. These checks have been made so far for four different situations, as mentioned above, characterized arbitrarily as "very moist" (28. 9.), "moist" (30. 10.), "dry" (11. 8.) and "very dry" (13. 8.). Fig. 4a—d shows the tensions, the joining lines representing the gradients of the matrix potential. The crosses in the r.h.s. column indicate the significance of the differences of the mean values between beech and spruce, and the crosses attached

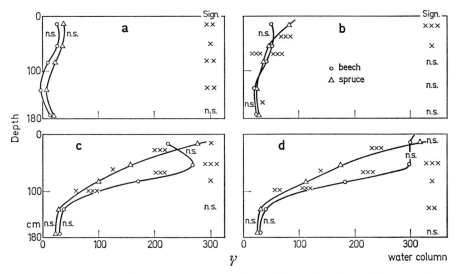

Fig. 4a—d. Gradients of the matrix potential ($\Delta\psi/\Delta h$) under four different moisture conditions

to the joining lines show the significance of differences between the various depths, "n. s." means "not significant" and is interpreted as "no difference in matrix potential". Though the lines are self-explanatory, it may be emphasized that the spruce site shows slightly drier conditions under "very moist", drier conditions at least in the top and bottom layer of the profile under "moist", whereas the beech apparently is able to make better use of the water when tension rises. Note that the whole range of tensions lies between about zero and 300 cm water column, i.e. less than 0.3 atmospheres. This result indicates that obviously under the given conditions spruce is equipped to deal with rather wet conditions without difficulty, whereas beech seems to prefer drier stands.

Another feature may be better seen in Fig. 5a—d which shows the hydraulic gradients computed as the sum of matrix and gravitational potentials. The ordinates show the distance from an arbitrarily chosen reference level and the abscissa only serves to read off the potential differences, the absolute numbers being meaningless. Since water movement is assumed to follow Darcy's law: $V = k \cdot \text{grad } \phi$ (i. e. the rate of flow equals the hydraulic gradient times the permeability coefficient), $\phi = \text{const}$ or $\text{grad } \phi = \dfrac{\Delta\phi}{\Delta h} = 0$ represented by a vertical straight line indicates "no potential difference" and consequently no water movement. Deviation to the left means upward and deviation to the right downward water movement. Fig. 5a shows approximately grad $\phi = 1$ indicating that an infiltration process governs the situation under both spruce and beech. As already mentioned, spruce seems able to maintain drier conditions under high precipitations. Fig. 5a and 5b show that this holds except for "very moist", where obviously infiltration governs the situation rather than withdrawal by the roots.

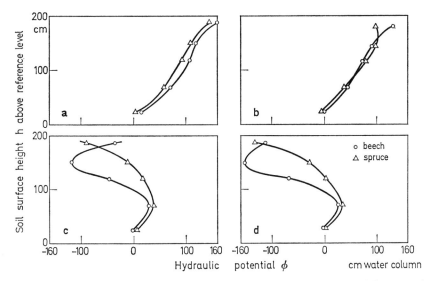

Fig. 5a—d. Gradients of the hydraulic potential ($\Delta\phi/\Delta h$) under four different moisture conditions

Fig. 5b shows how spruce takes its main water supply from the top layer, whereas beech achieves an almost parallel shift of the whole line into the drier region. As may be concluded from Fig. 5c, there seem to be three simultaneous processes involved: withdrawal by roots, mainly at a depth about 60 cm below surface, i.e. in the lower part of the loess layer; supply of this region from the top layer and later on from the layer down to about 120 cm; and eventually permanent downward seepage through the subsoil into the underlying permeable "Buntsandstein". A comparison of Fig. 5c and 5d stresses the different behavior of spruce and beech with respect to water use and water withdrawal.

## V. Flow Rates — a First Approximation

Going back to Fig. 4a—d, no differences can be observed between spruce and beech in the subsoil (120—180 cm) under medium conditions but significant differences are seen under "very moist" and "dry" conditions, the "beech-gradient" lying in a region of higher tensions under "dry" and of lower tensions under "very moist".

To obtain the corresponding flow rates, information about the permeability is needed. Measurements have been made on undisturbed soil cores cut out with

stainless-steel cylinders of 250 cm³ volume (Henseler and Renger, 1969) as well
as on large undisturbed soil cores of 125 cm length and 28 cm diameter. Since interest
is focussed mainly on the layers below the root zone in order to compute the
amount of groundwater recharge, only measurements at this depth will be considered. The results shown in Fig. 6 should be regarded as preliminary only. Further
confirmation is necessary, and it is intended to check the permeability data by obtaining them by various methods; it may be pointed out, however, that the permeability data shown in Fig. 6 were yielded by quite different methods but nevertheless
are in good agreement.

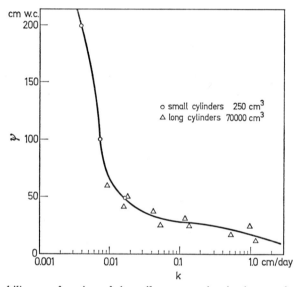

Fig. 6. Permeability as a function of the soil water tension in the root-free subsoil horizon

As a first approximation, the results shown in Fig. 6 may give some idea of the
probable rate of flow and the difference between spruce and beech. Fig. 4a—d shows
that in general the soil water tension of the subsoil lies within a range about 30 cm
water column. On the other hand the relation $k$ versus $\psi$ (the matrix potential or
tension) in Fig. 6 shows almost a plateau in this region, indicating that the permeability is very sensitive to even small changes in tension. Since the hydraulic gradient is
in general close to 1, we may (cf. Darcy's law) directly equate the flow rate with the
permeability coefficient. Now it will be seen at once that a substantial amount of
groundwater recharge is possible only if $k$-values are about 0.1 cm/day, which
requires tensions below 30 cm water column. Obviously, we have to conclude from
the data that, except under the condition "very moist", there is little groundwater
recharge, but that under "very moist" conditions substantial flow rates may occur.
    Taking this into consideration, the significant difference between the "beech-
gradient" and the "spruce-gradient" becomes very meaningful, indicating that during
the crucial periods the groundwater recharge under beech may well be twice as high
as under spruce.

# References

Arbeitsgemeinschaft Bodenkunde: Die Bodenkarte 1:25000. Hannover: Niedersächsisches Landesamt für Bodenforschung 1965.

Brülhart, A.: Jahreszeitliche Veränderungen der Wasserbindung und der Wasserbewegung in Waldböden des schweizerischen Mittellandes. Mitt. Schweiz. Anst. f. d. Forstl. Versuchswesen 45, 127—232 (1969).

Eschner, A. R.: Interception and soil moisture distribution. In: Forest Hydrology, Symposium held at the Pennsylvania State University, Aug./Sept. 1965, pp. 191—200. Oxford: Pergamon Press 1965.

Hartge, K. H., Benecke, P.: Wasserspannungsgradienten in einem Schluff- und einem Sandboden. Wasser u. Boden 1, 18—21 (1968).

Harvey, W. R.: Least-squares analysis of data with unequal subclass numbers. Agric. Res. Serv. U. S. Dept. of Agriculture, 20, 8 (1960).

Henseler, K. L., Renger, M.: Die Bestimmung der Wasserdurchlässigkeit im wassergesättigten Boden mit der Doppelmembran-Druckapparatur. Z. Pflanzenernähr., Düngung u. Bodenkunde 122, 220—228 (1969).

Rose, C. W.: Agricultural Physics, Oxford: Pergamon Press 1966.

Swartzendruber, D.: Soil-Water behavior as described by transport coefficients and functions. Advanc. Agron. 18, 327—370 (1966).

# Va. Methods of Investigating the Micromorphology of Humus*

U. BABEL

## I. Introduction

The study of humus morphology is based on the work of P. E. MÜLLER (1887) and HESSELMAN (1926) in its macromorphological part, and that of KUBIENA (1943, 1953) in its micromorphological part. The subject of humus morphology is the humus profile, that is, the part of the soil profile whose properties are determined by organisms and dead organic substances.

A short review of methods is given here, especially regarding the investigation of moder humus forms, as present in the sample areas of the Solling Project. A more detailed presentation of the micromorphological methods can be found in BABEL (in press, a). For macromorphological descriptions of the humus profiles see BENECKE and MAYER, chapter U II.

## II. Description in the Field

The frame for the description of the humus profile in the field is given by the subdivision into horizons. The nomenclature of HESSELMAN is generally used, i.e. $L$, $F$, $H$ for the superficial humus layers; in Germany the upper humus horizon of the mineral soil is termed $A_h$ (*Arbeitsgemeinschaft Bodenkunde*, 1965). A finer subdivision is often possible and suitable when more detailed investigations are required, despite the fact that some of these subhorizons are only a few millimeters thick (for details see BABEL, in press, b).

$L$-layer: litter of plant cover, with no organic fine substance or very little (about 10% at most).

$L_n$ (n from lat. novus, new) litter not clearly changed morphologically as compared with the state immediately after death.

$L_v$ (v from German „verändert") litter clearly changed (mechanically disintegrated, changed in colour, bound by fungus hyphae etc.)
Note: This subhorizon often shows marked changes due to microbial attack and removal of organic substances; macromorphologically, however, it is still closer to $L_n$ than to $F_r$, as the amount of fine substance is still very low.

$F$-layer, fermentation layer: plant residues easily recognizable with the naked eye, low or medium contents of organic fine substance.

$F_r$ (r from plant residues) plant residues prevailing (about 10—30% organic fine substance).
Note: This is the horizon where the activity of litter-disintegrating animals prevails.

$F_m$ (m from medium) medium state of litter disintegration; medium amounts of plant residues as well as of fine substance (about 30—70% organic fine substance).

---

* See also KUBIENA and GEYGER, chapter Vb.

*H*-layer, humus substance layer: few or no residues of above-ground plant organs, organic fine substance prevailing, mineral substance 50 vol.% at most (= about 65 weight %).

$H_r$ (r from plant residue) residues of above-ground plant organs to be found easily, without seeking (about 10—30%); root residues not taken into account.

$H_f$ (f from fine substance) residues of above-ground plant organs are recognizable only after close investigation or are completely absent (10% at most); root residues not taken into account.

$A_h$-horizon: uppermost mineral soil horizon, clearly coloured by humus.

$A_{hh}$ (doubling of h because of high humus content) uppermost part of the $A_h$, strongly to very strongly coloured by humus.

Note: This subhorizon is clearly seen in nearly all moder profiles; its thickness amounts to 1—2 cm.

$A_{hu}$ (u for subhorizon under the $A_{hh}$) horizon of the upper mineral soil, medium to weakly coloured by humus, with medium dense or dense "fabric" (see section V).

The features of the macromorphological description of the humus profile are estimated on a scale from 0—5 for amount or intensity. To characterize the fabric in superficial humus layers, it is usually sufficient to describe the features of coherence (degree of coherence of particles) and density (complementary to the macroscopically visible pores). Droppings of large arthropods and their larvae can easily be seen with the naked eye. To describe the decomposition of plant residues, the degree of mechanical disintegration of the leaves, the discolouration (by fungi) and the sceletonization (by animals) are recorded.

## III. Investigations by Stereomicroscope

The macroscopic and microscopic investigations are supplemented by the stereomicroscope. Observations can be made on the first stages of decomposition of leaves (e.g. separation of the epidermis) and on fungal attack. Small droppings (of mites, Enchytraeidae and others) can be seen. The smaller soil animals can be found easily by opening aggregates with preparation needles. Thus a rough idea can be obtained of the soil animals of the groups of Enchytraeidae, Oribatid mites, Collembolae and small insect larvae present at the moment of sampling.

The stereomicroscope serves to isolate plant residues which are to be prepared for cutting into thin sections.

Stereomicroscopic investigations of polished sections (soil samples which are embedded in artificial resin, sectioned and polished at one edge only, and covered with glycerine or any other highly refractive liquid) serve to investigate the microfabric. If not thinner than 20 $\mu$, ground thin sections may also be investigated with the stereomicroscope. This often reveals details of the microfabric better than the microscope; this is especially true for enchytraeidae droppings (BABEL, 1967).

## IV. Technique of Investigating Ground Thin Sections

### 1. Sampling and Preparation

The samples are taken without disturbing the fabric (e.g. in metal frames). Thin sections of humus soil horizons are produced by the techniques generally applied to mineral soils — embedding in polyester resin after air drying — (ALTEMÜLLER, 1956 and 1962; BABEL, in press, a; BREWER, 1964). The artefacts arising from this preparation are insignificant (except many phenomena of shrinking) and generally need not be taken into consideration.

## 2. Microscopic Techniques

### a) Morphometry

Besides counting and measuring length in thin sections, volume ratios can be determined by the point counting method (Zeiss point counter eyepiece — for fundamentals see Hennig, 1958). For comparative investigations, estimations of quantities are often suitable (scale 0—5, Babel, 1967). Methods and applications of micromorphometric soil analysis are treated in detail by Kubiena (1967); see also chapter Vb.

### b) Chemical Techniques

Simple chemical tests can be done on the ground thin section (Babel, 1964b). Of special interest is the treatment with sodium hypochlorite (Eau de Javelle, NaOCl), which bleaches the organic substance due to oxidation of humic substances. This method enables mineral and organic particles to be distinguished in doubtful cases; in addition, tissue structures in strongly coloured organic particles become visible.

### c) Optical Techniques

Illumination with phase contrast, dark field and incident light are of importance in special cases only. Of general importance is the investigation of humus thin sections with the polarizing microscope. The strong positive birefringence is a good test reaction for cellulose. In mineral humus horizons colourless mineral grains are made visible by their birefringence. Investigation with the fluorescence microscope has a similar effect: nearly all cell walls as well as minute cell wall residues become visible and, when primary fluorescence in UV-light is used, are contrasted from their environment much better than by birefringence. The primary fluorescence of the embedding medium is generally very low and does not interfere.

# V. Cut Thin Section Study and Debris Preparation

## 1. Sampling and Preparation

Superficial humus layers are prepared carefully, layer by layer, subdividing them into as many subhorizons as possible. — Cut thin sections are prepared from plant residues when details of their disintegration are to be investigated histologically. Paraffin wax thin sections may be used according to the usual botanical techniques (e.g. Johansen, 1940); fixation is not required. — Debris preparations serve for the histological investigation of the organic fine substance. The fine substance is prepared by wet sieving at 100 or 200 $\mu$.

## 2. Microscopic Techniques

Besides the techniques applicable to ground thin sections, the test reactions for plant substances usual in botany may be applied to cut thin sections and debris preparations: Sudan III for cutin, phloroglucinol for lignin, solution of iodine in zinc chloride for cellulose, ruthenium red for pectin. These colour reactions, however, can be applied only to colourless or weakly coloured material. The coloured humified substance may be removed by a pretreatment with sodium hypochlorite; this pretreatment should be applied with great care (e.g. dilution 1:4, that is 2.5%, 5 min).

# VI. Particle and Fabric Analysis

The following techniques are used for particle and fabric analysis.

## 1. Particle Analysis

Soil organic matter may be subdivided micromorphologically into 3 groups of particles: plant residues (particles with easily recognizable structures or contours of tissue, greater than about 100 $\mu$), organic fine substance (purely organic particles which contain only very small cell groups or cell residues, if any), organic substance as pigment of mineral fine substance.

The aims of particles analysis are: (1) Identification of plant residues; as regards disintegration, they may be subdivided into residues of parenchyma, of lignified tissues, and of phlobaphene-containing tissues (BABEL, 1965); and (2) morphological description, comprising transformation state of tissue structure, preservation of birefringent cell walls, occurrence of coloured substances in the tissue residues. These investigations may be done on ground thin sections. Systematic investigations of the changes in the tissue and cell parts in the course of decomposition are done on cut thin sections.

## 2. Fabric Analysis

Fabric analysis concerns the arrangement of the particles (elementary fabric) and size, form and arrangement of aggregates and pores (aggregate fabric). Therefore fabric analysis is always done on ground thin sections. Important features are: the arrangement of plant residues in relation to organic fine substance and mineral substance (e.g. leaf residues blended into the mineral soil by earthworms), also the blending of organic fine substance into the mineral soil. Important processes of aggregation are the formation of leaf packets and the binding of plant residues and aggregates by bridges of fine substance or by hyphae.

The most important aggregates in the humus profile are excrements of soil animals ("droppings" or "pellets") and their transformation products. Different groups of soil animals have characteristic forms of droppings (see ZACHARIAE, 1965). The localization, colour and structure of the droppings give clues to the food of the animals.

# VII. Application of Humus Micromorphology

Work with these investigational methods gives results in the following range of problems:

Soil physics: Size and form of aggregations and pores and their formation and transformation by abiotic and biotic factors (shrinking, surface tension of the water; roots, animals). Aeration and air moisture in the humus profile and at the micro-habitats for the soil organisms may be estimated.

Humus chemistry: Stability of the plant substances, their relative rates of degradation, new compounds formed topochemically from the plant substances, formation of organic particles by precipitation from the soil solution.

Soil biology: Microhabitats of animals and fungi; food, disintegration and transport activities of soil animals; role of fungi in decomposition and formation of relatively stable humus substances.

The investigation of humus morphology attempts to find and explain the qualitative correlations in the formation and processing of the humus profile.

# References

Altemüller, H. J.: Neue Möglichkeiten zur Herstellung von Bodendünnschliffen. Z. Pflanzenernähr., Düngung, Bodenkunde 72, 56—62 (1956).
— Verbesserung der Einbettungs- und Schleiftechnik bei der Herstellung von Bodendünnschliffen mit Vestopal. Z. Pflanzenernähr., Düngung, Bodenkunde 99, 164—177 (1962).
Arbeitsgemeinschaft Bodenkunde: Die Bodenkarte 1:25000. Hannover: Niedersächsisches Landesamt für Bodenforschung 1965.
Babel, U.: Dünnschnittuntersuchungen über den Abbau lignifizierter Gewebe im Boden, S. 15—22. In: Jongerius, A. (ed.): Soil micromorphology. Proc. 2nd Int. Working Meeting on Soil Micromorphology, Arnhem, Amsterdam: Elsevier 1964a.
— Chemische Reaktionen an Bodendünnschliffen. Leitz-Mitt. 3 (1), 12—14 (1964b).
— Vergleich von Mikrogefügemerkmalen einiger Humusbildungen mit Hilfe einer Schätzmethode. Geoderma 1, 347—357 (1967).
— Micromorphology of soil organic matter. Encyclopedia of Soil Science. (Ed. J. E. Gieseking), Vol. IIC (in print, a).
— Gliederung und Beschreibung des Humusprofils in mitteleuropäischen Wäldern. Geoderma (in print, b).
— Fluoreszenzmikroskopie in der Humusmikromorphologie. 3rd Int. Working Meeting on Soil Micromorphology, Sept. 1969, Wroclaw, Poland (in print, c).
Brewer, R.: Fabric and mineral analysis of soils. London-New York: Wiley and Sons 1964.
Hennig, A.: Kritische Betrachtung zur Volumen- und Oberflächenmessung in der Mikroskopie. Zeiß-Werkzeitschrift 30, 78—86 (1958).
Hesselman, H.: Studier över barrskogens humustäcke. Meddel. Stat. Skogsförsöksanst. 22, 169—552 (1926).
Johansen, D. A.: Plant microtechnique. New York: Mc Graw Hill 1940.
Kubiena, W.: Die mikroskopische Humusuntersuchung. Z. Weltforstwirtschaft 10, 387—410 (1943).
— Bestimmungsbuch und Systematik der Böden Europas. 392 pp. Stuttgart: Enke 1953.
— (Ed.): Die mikromorphometrische Bodenanalyse. Stuttgart: Enke 1967.
Müller, P. E.: Studien über die natürlichen Humusformen. Berlin: Springer 1887.
Zachariae, G.: Spuren tierischer Tätigkeit im Boden des Buchenwaldes. Forstwiss. Forschungen (Beih. Forstl. Cbl., 20 (1965).

# Vb. Methods Used in Micromorphological and Micromorphometric Soil Studies*

W. L. KUBIENA† and E. GEYGER

## I. Introduction

The microstructure of the soil and, particularly, of its pore system is highly important for the growth and production of higher plants, micro-organisms, and many animals. The "micromorphology" of the soil provides exact information about these properties through the use of microscopic methods to investigate the material and structural combination of the individual elements in naturally deposited soil (KUBIENA, 1931, 1937, 1938, 1941, 1962, 1964). Subsequently, a quantitative research technique, "micromorphometry", (KUBIENA, BECKMANN and GEYGER, 1961—1963; BECKMANN, 1962, 1964; GEYGER, 1962; GEYGER and BECKMANN, 1967) has been developed on the basis of polished soil sections and thin sections. This technique provides a means of precisely determining the elements of soil structure, such as pores and aggregates, in their natural diversity of shape, as well as their relationships of size, amount and position. In addition, these methods make it possible to measure and describe changes of the structure occurring in the course of a year or as a result of agricultural management.

## II. Freeze-Drying Methods

Very recently, methods of freeze-drying have been put to use for the production of soil thin sections in order to be able to investigate the soil structure in its naturally moist condition (STEPHAN, 1969, here also references to earlier literature): The freshly extracted, naturally moist soil samples are rapidly deep-frozen by dipping into liquid nitrogen, thus avoiding the formation of large ice crystals which are capable of disrupting the natural soil fabric. Drying ensues in a vacuum at temperatures far below freezing point, so that ice evaporates immediately. In this way, shrinkage due to loss of water, a disturbing factor in connection with air-drying, is eliminated; this freeze-drying process rarely results in shrinkage of the soil samples. The spatial relationships of the soil particles are largely retained as they existed in their naturally moist state; the original size relationships of the plant structures are preserved as well.

This has special significance for soils which, in their natural state, are permanently moist, as is the case for the majority of the Solling sites. We are now much better equipped to use the same soil samples whose physical properties (porosity, pF) were studied by BENECKE and MAYER (chapter U) to ascertain micromorphometric data as well, such as form, size and volume percentage of different types of soil pores. It is

---

* See also BABEL, chapter Va.

expected that this new combination of methods will allow insights into the relationship between the physical and morphological characteristics of the soil, especially with respect to the water conditions of the soil. Since work with this technique has only just begun, results cannot yet be presented. Freeze-drying also facilitates the study of soil structure as a mirror of soil-biological effects. Our contribution to this cooperative work lies in identifying pellets from various groups of animals, and in determining their respective shares in the composition of the humus horizons, as well as in determining the structural changes undergone by the litter in the course of its gradual disintegration. Thus, a highly precise morphological characterization of the humus form can be attained.

# References

Beckmann, W.: Zur Mikromorphometrie von Hohlräumen und Aggregaten im Boden. Z. Pflanzenernähr., Düngung, Bodenkunde 99, 129—139 (1962).
— Zur Ermittlung des dreidimensionalen Aufbaues der Bodenstruktur mit Hilfe mikromorphometrischer Methoden. In: Soil Micromorphology, ed. by A. Jongerius, Amsterdam, 429—444 (1964).
Geyger, E.: Zur Methodik der mikromorphometrischen Bodenuntersuchung. Z. Pflanzenernähr., Düngung, Bodenkunde 99, 118—129 (1962).
Geyger, E., Beckmann, W.: Apparate und Methoden der mikromorphometrischen Strukturanalyse des Bodens. In: Die mikromorphometrische Bodenanalyse, ed. by W. Kubiena, Stuttgart, 36—57 (1967).
Kubiena, W. L.: Mikropedologische Studien. Arch. f. Pflanzenbau Abt. A, 5, 613—648 (1931).
— Verfahren zur Herstellung von Dünnschliffen von Böden in ungestörter Lagerung. Zeiss Nachrichten 2, 3, 1—11 (1937).
— Micropedology, Ames (Iowa): Colleg. Press Inc. 1938.
— Die Dünnschlifftechnik in der Bodenuntersuchung. Der Forschungsdienst, Sonderheft 16 (1941).
— Wesen, Ziele und Anwendungsgebiete der mikromorphologischen Bodenforschung. Z. Pflanzenernähr., Düngung, Bodenkunde 97, 193—205 (1962).
— The role and mission of micromorphology and microscopic biology in modern soil science. In: Soil Micromorphilogy, ed. by A. Jongerius, Amsterdam, 1—13 (1964).
— Beckmann, W., Geyger, E.: Zur Methodik der photogrammetrischen Strukturanalyse des Bodens. Z. Pflanzenernähr., Düngung, Bodenkunde 92, 116—126 (1961).
— — — Zur Untersuchung der Feinstruktur von Bodenaggregaten mit Hilfe von Strukturphotogrammen. Zeiss-Mitt. 2, 256—273 (1962).
— — — Mikromorphometrische Untersuchungen an Hohlräumen im Boden. Anal. Edafolog. y Agrobiol. 22, 551—567 (1963).
Stephan, S.: Gefriertrocknung und andere bei der Herstellung von Bodendünnschliffen benutzbare Trocknungsverfahren. Z. Pflanzenernährung und Bodenkunde 123, 131—140 (1969).

# W. Soil Chemical Differences between Beech and Spruce Sites — an Example of the Methods Used

B. Ulrich, E. Ahrens and M. Ulrich

## I. Introduction

### 1. General Framework

The progress of a science depends upon the suitability of the methods available for studying the problem in question. The methods used in soil chemistry have been developed either for characterizing the soil as a natural body or for indicating fertilizer needs in agricultural practice. The latter methods are more or less well suited for their special purpose but they yield no information of scientific significance. These methods therefore should not be used for characterizing the soil as part of an eco-system. The methods of characterizing the soil as a natural body yield information about the chemical properties of the soil and the processes going on in it, but these are not necessarily ecologically significant properties and processes. Since no specific ecopedological methods have been developed so far, the methods developed in the field of soil chemistry must be used in the field of ecopedology, too; but they have no immediate ecological meaning — this must be derived by comparison.

In the frame of the Solling Project the chemical soil characterization will be used in combination with the following investigations:

Bioelement inventory: In the bioelement inventory the amount of bio-elements in soil and vegetation cover, both divided into appropriate strata, are considered. The amount of bioelements in the soil, present in specific binding forms, represents extensive properties of the soil, which depend on soil mass: doubling the mass yields double the amount of extensive properties. In soil sampling, volume samples should therefore be taken, or the volume weight of the soil determined separately. Since the extensive properties of a system are additively composed of the same properties of the various parts of the system, a bioelement inventory of a site can be made up if, for all strata considered, the amounts of bioelements are expressed per area.

Bioelement cycle: The bioelement cycle of an ecosystem consists of the following main processes:

a) Input from the atmosphere by precipitation, dust etc.
b) Uptake by the vegetation cover from soil (and atmosphere)
c) Transport within individual plants
d) Transport from the vegetation cover to the soil by litter fall, crown washout and stem flow
e) Transport within the soil
f) Output to the ground water (and atmosphere)

Of these processes, a, d and f are investigated by measuring the fluxes of transport medium (precipitation, litter, soil water) and determining the bioelement content in the transport medium. In connection with the bioelement inventory the investigations are intended to allow an approximate calculation of process b. Further information is expected from a comparison of the bioelement uptake by the vegetation cover with the bioelement stores in the soil and the bioelement composition of the soil solution.

For the transport processes occurring within the soil, a set of suction lysimeters has been installed at various soil depths; the solution is collected and its bioelement content determined. These data and the data characterizing the equilibria between solid soil phase and equilibrium soil solution (Schofield's potentials, Gapon coefficients, solubility products) will be used together with literature data about diffusion coefficients of bioelements in soil to set up models for the description of the transport processes within the soil.

The methodical questions involved in investigations regarding bioelement inventory and bioelement cycling are mainly those of sampling techniques and sampling errors; they will be treated in a later publication.

In the following the methods used for the characterization of chemical soil properties are demonstrated by a comparison of the influence of beech vs. spruce on the soil.

## 2. Research Area

In the Solling region the natural vegetation corresponds to a beech forest *(Luzulo-Fagetum)*. In earlier centuries parts of the beech forests were devastated by man, and for the reafforestation in the mid-19th century mainly spruce *(Picea abies)* was used. Since the same history is repeated in large parts of Central Europe, the influence of spruce versus beech on the soil has long been under discussion.

Here the results of chemical soil investigations of three beech and three spruce stands in the Solling are presented. The stands were selected after soil mapping, so only soils of the same parent materials and type are included. A detailed description of the soil type is given by Benecke and Mayer (chapter U). The parent materials are a loess creeping earth of 40—60 cm thickness above a para-autochthonic creeping earth consisting of loamy weathered Buntsandstein material. The unweathered Buntsandstein of Triassic age begins at depths of about 100 cm. There is naturally

Table 1. *Characterization of the plots*

| name of plot | B 1 | B 3 | B 4 | F 2 | F 3 | F 4 |
|---|---|---|---|---|---|---|
| tree species | beech *(Fagus silvatica)* | | | Norway spruce *(Picea excelsa)* | | |
| age of stand (yrs) | 120 | 78 | 57 | 113 | 39 | 18 |
| soil type | acid brown earth (typic dystrochrept) | | | | | |
| humus form | moder | moder | mull-like moder | thick moder | moder | moder |
| podzolic tendency | weak | strong | weak | strong | medium | medium |

Plot F 1 is not included. This plot was selected later and soil sampling was therefore done in spring and not in autumn as for the other plots. Since the time of sampling influences some of the soil solution data, the data for F 1 will be reported in a later paper in context with this question. (Plot F 4 was investigated in addition to the general programme).

some variation within the plots selected as well as between the plots, mainly in respect to the thickness of the loess creeping earth and to the kind of Buntsandstein creeping earth, which is sandier in some parts. Since the loamy creeping earth has a dense structure and hence a low water permeability, there is also some variation in respect to waterlogging, in certain areas the lower part of the loess creeping earth may show some bleaching and a few small iron concretions. According to the German nomenclature the soil is an acid brown earth (typic dystrochrept of the 7th approximation). A description of the humus forms and podzolic tendencies in the $A_h$ horizon is summarized in Table 1.

# II. Soil Chemical Methods Used

## 1. Bioelement Stores

From the ecological point of view the amount of bioelements in a soil (the capacity terms) should be defined in relation to the possibility of turnover. The amount of bioelements having the best chance of entering the bioelement cycle may be called the mobilizable bioelement pool. In principle one can assume that all bioelements taken up by plants and stored in plant cover or in the top organic layer of forest soils, as determined by ash analysis, have already been subjected to turnover and thus belong in total to the mobilizable pool, irrespective of the binding form. For the bioelements in soil, on the other hand, a subdivision is necessary, leaving out binding forms with commonly very low turnover rates from the mobilizable pool. In the following account, the index $t$ indicates total amounts and the index $m$ indicates mobilizable amounts.

### a) Carbon and Nitrogen

In the case of carbon and nitrogen no subdivision has been carried out: only $C_t$ and $N_t$ in soil and the C/N ratio have been determined by common analytical methods.

### b) Phosphorus

In the case of phosphorus in mineral soil, the fractionation procedure of CHANG and JACKSON (1957) has been adopted in the form published by KHANNA et al. (1967): $NH_4F$ extraction (Al-phosphates $P_{Al}$), followed by NaOH extraction (Fe-phosphates, $P_{Fe}$), followed by $H_2SO_4$ extraction (in the case of strongly acid soils as in the Solling: acid-soluble occluded phosphates $P_{a.s.}$), followed by total P determination (occluded phosphates $P_{occl}$). The difference between $P_t$ in the original soil and the sum of all fractions yields the organic bound phosphorus ($P_{org}$), which includes therefore all errors. The sum of $P_{Al}$, $P_{Fe}$ and $P_{org}$ constitutes the mobilizable phosphate pool.

### c) Cations

In the case of *cations* (H, Na, K, Ca, Mg, Al, Fe and Mn) in mineral soil, only the exchangeable cations (index $e$) can be attributed to the mobilizable pool. They have been determined by leaching the soil with 1 n $NH_4Cl$ according to ULRICH (1966a). By this method, in acid soils the exchange of cations including H takes place at nearly the same pH value as in the natural soil. The sum of all exchangeable cations mentioned above is called effective cation exchange capacity ($CEC_e$).

### d) Total Cation Exchange Capacity

The total cation exchange capacity $CEC_t$ has been determined with $NH_4$-oxalate in the presence of $CaCO_3$ according to Riehm et al. (1954). In soils rich in organic matter, as in the top mineral soil layers of acid forests soils, this rapid method has the disadvantage of dissolving some organic matter, so yielding too low a value for $CEC_t$.

The difference between $CEC_t$ and $CEC_e$ represents the *pH-dependent* CEC, which is due to either undissociated OH groups or the blocking of permanent charges by Al polymers or allophane.

The analytically determined values can be expressed as contents (e. g. % C, % N, mg P/g soil, $\mu$eq exchangeable cations/g soil, $\mu$eq CEC/g soil), or as stores (tons, kg, keq per ha) or as fractions. In the last case one gets dimensionless ratios, which are well suited to compare soils different in texture, organic matter etc. in relation to their state properties. The following fractions will be used (*i* indicates any of the cations mentioned before)

$X_i^S$: cation equivalent fraction in exchangeable cations, calculated by dividing the content of cation *i* by $CEC_e$.

Other dimensionless ratios need no specification, like $C : N$, $C : P_{org}$, $CEC_e : CEC_t$. These ratios do not represent extensive properties of the soil as the stores do, but intensive properties which are independent of soil mass.

## 2. Equilibrium Soil Solution and Schofield's Potentials

The soil solution represents that constituent of the soil with which the plant root is in closest contact or which may even enter the outer space of plant roots. Comparing the bioelement composition of soil solution and of plant uptake enables conclusions to be drawn about selectivity of bioelement uptake (Ulrich, 1966b). For extraction of equilibrium soil solution (ESS) several procedures can be used. Usually before extractions water is added to the dry or field-moist soil, and after equilibration is extracted again by displacement with other liquids (Moss, 1963), pressure filtration (le Roux et al., 1967), centrifugation or vacuum filtration. The added water should not exceed 0.6—0.8 ml/g soil. In the work described here the ESS was extracted by vacuum filtration of a saturated soil slurry (saturation extract according to Richards, 1954), starting from a soil volume of 600 $cm^3$. The ions H, Na, K, Ca, Mg, Al, Fe, P, S and Cl in the soil solution were determined by conventional methods.

The composition of the soil solution is expressed as a concentration $c$ ($\mu$g-atom/l) or as a dimensionless ratio (*i* indicates any of the cations mentioned) $X_i^L$: cation equivalent fraction in ESS, calculated by dividing the equivalent concentration of cation *i* ($\mu$eq/l) by the equivalent sum of all cation concentrations in soil solution. Whereas the ion concentrations in ESS strongly depend upon the water:soil ratio in the saturated soil slurry (Ulrich et al., 1970), the same is true to a lesser degree of the $X^L$ values. These $X^L$ values are therefore more convenient for comparing different soils in respect to composition of ESS.

Expressions of the composition of ESS which are independent of electrolyte concentration and the water:soil ratio used for extracting ESS represent Schofield's potentials (*SP*). For cations a Schofield's potential is defined by

$$SP_{A/B} = 1/z_A\, pA - 1/z_B\, pB$$

where $A$ and $B$ are two cations with the valency $z_A$ resp. $z_B$, and $p$ indicates the negative logarithm of the ion activity of the ions $A$ and $B$. The ion activity has been calculated from the ion concentration $c$ by

$$a_i = f_i \cdot c_i$$

where $f$, the activity coefficient, has been calculated according to the Debye-Hückel equation

$$-\log f_i = (0.51 \cdot z_i^2 \sqrt{\bar{I}})/(1 + \sqrt{\bar{I}})$$

I, the ionic strength, is calculated by

$$I = 1/2 \sum_i c_i \cdot z_i^2$$

## 3. Exchange and Solubility Equilibria

The analysis of exchangeable cations as well as of the ESS allows the characterization of cation exchange equilibria by means of the Gapon equation

$$A_s/B_s = k^G_{A/B} \cdot R_{A/B}$$

$R_{A/B}$, the so-called reduced ratio, is defined by

$$R_{A/B} = z_A \frac{}{\sqrt{a_A}} \Big/ z_B \frac{}{\sqrt{a_B}}$$

Therefore $SP_{A/B} = -\log R_{A/B}$

The Gapon coefficient $k^G_{A/B}$ is a measure of the selectivity of cation binding at cation exchanger surfaces.

From the cation concentration in the ESS, solubility products can also be calculated. As far as solubility products are used, they will be described in the context.

From Schofield's potentials measured at different soil depths a chemical potential $\mu_i$ for any cation i may be calculated, describing the tendency for cation exchange reactions during the transport of soil solution through the soil profile (ULRICH, 1969a). Calculation and use of these data is described later.

# III. Sampling Procedures and Statistical Treatment of Data

From each of the 6 plots with an area of at least 0.5 ha sampling has been twofold (sampling replicates $d_1$ and $d_2$, Table 2). Each of the replicate samples is a mixture of 6 subsamples, collected according to a regular network distribution over the whole area. In the places where the subsamples were to be taken the organic top layer (O horizon), separated into $L$, $F$ and $H$ layers, was collected quantitatively by means of a metal cylinder with an area of 0.5 m². In the same places the soil-boring tool described by ATANASIU et al. (1961) was used to collect volume samples of mineral soil at 10 cm depth intervals down to 50 cm depth. For the statistical treatment, the values for the $L$, $F$ and $H$ layers have been combined, so for most variables six depth intervals will be considered.

For the statistical treatment the analysis of variance has been used (cf. Table 2). The effects of tree species ($A$), of plots within tree species [$B(A)$] and of sampling within plots within tree species [$P(B)(A)$] are considered to be random, for it is

Table 2. *Sampling scheme and symbols used in analysis of variance*

| Effect (factor) | | Index | Ranks of factor | Sampling scheme | | Designation of ranks | Hierarchic steps | Type of effect | Designation of main effect for each rank |
|---|---|---|---|---|---|---|---|---|---|
| | | | | $a_1$ beech | $a_2$ spruce | | | | |
| between tree species | $A$ | $i$ | $a = 2$ | | | $x_{i\cdots\cdots}$ | main group | random | $\alpha_i$ |
| plots within tree species | $B(A)$ | $j$ | $b = 3$ | $b_1\ b_2\ b_3$ | $b_1\ b_2\ b_3$ | $x_{ij\cdots\cdots}$ | 1st sub-group | random | $\beta_{ij}$ |
| sampling within plots within trees | $P(B)(A)$ | $l$ | $d = 2$ | $d_1\ d_2\ d_1\ d_2\ d_1\ d_2$ | $d_1\ d_2$ | $x_{ij\cdot l\cdot}$ | 2nd sub-group | random | $\delta_{ijl}$ |
| between depth | $C$ | $k$ | $c = 4$ or 5 or 6 | $c_1\ c_1\ c_1\ c_1\ c_1\ c_1$ | $c_1\ c_1\ c_1\ c_1$ | $x_{\cdots k\cdots}$ | main group | fix | $\gamma_k$ |
| | | | | $c_2\ c_2$ | | | | | |
| analytical replicates | $R$ | $m$ | $e = 2$ | $c_3$ $c_4$ $c_5$ | | $x_{ijklm}$ | sub-group random | | $\varepsilon_{ijklm}$ |

the object of the investigation to draw inferences about the population of soils under beech and spruce. On the other hand, the soils selected reflect only a small section of the soil population under both tree species. If soils differ in their chemical properties, e.g. in pH, they usually differ in the depth functions of these properties. Therefore the effect depth $(C)$ is considered fixed, that is, we can draw inferences only about the particular depth functions reflected in the soils examined.

The statistical model is therefore a mixed one, not only due to the occurrence of random and fixed effects, but also due to the combination of a two-way classification with hierarchic steps, as shown in Table 2. It can be described by the following equation where $y_{ijklm}$ is the measured value of any variable at any place in the sampling scheme of Table 3:

$$y_{ijklm} = \mu + \alpha_i + \beta_{ij} + \delta_{ijl} + \gamma_k + (\alpha\gamma)_{ik} + (\beta\gamma)_{ijk} + (\delta\gamma)_{ijkl} + \varepsilon_{ijklm}$$

In this equation, $(\alpha\gamma)_{ik}$ corresponds to the interaction $A \times C$, $(\beta\gamma)_{ijk}$ to the interaction $B(A) \times C$ and $(\delta\gamma)_{ijkl}$ to the interaction $P(B)(A) \times C$. Since the effects can be considered as a linear combination of the kind expressed in the equation, the analysis of variance can be used to evaluate the data.

For further evaluation of effects, the method of partitioning degrees of freedom and sums of squares has been used as described by STEEL and TORRIE (1960) in order to obtain sets of contrast, especially for the effect $A$ (tree species) within the different soil depth intervals.

It should be mentioned that a test of the type of distribution did not show normal distribution for all variables. A one-sided distribution was found for the following variables (the values represent the total variability and the peak of the distribution curve):

| | |
|---|---|
| pH (CaCl$_2$) | (3.1— 4.4; 4.1) |
| C/N | (9 — 30; 18) |
| exchangeable Ca | (0.8— 6.4; 1.4 $\mu$val/g) |
| exchangeable Mg | (0.2— 1.2; 0.4 $\mu$val/g) |
| exchangeable Fe | (0.1— 3.5; 0.25 $\mu$val/g) |
| exchangeable Mn | (0.4— 3.6; 0.7 $\mu$val/g) |
| H ion concentration in ESS | (20 —600; 80 $\mu$g-atoms/l). |

All one-sided distributions can be traced back to the influence of depth on the variables, mainly on the decrease in organic matter and soil acidity with increasing depth under the special conditions of an acid soil with an organic horizon on top. It is not thought that the deviation from normal distribution affects the analysis of variance of these variables to any great extent.

The figures and tables list mean values, components of variance (expressed as deviations from the mean $\bar{x}$ .....) and the significance of differences between means according to the $F$ test. The asterisks $*$, $**$ and $***$ (and crosses in Figs. 2—5) mean significance at the 5%, 1% and 0.5% level of probability, respectively, n.s. means not significant but is usually omitted. Significant effects mean:

Significance of A: significant differences in the mean values $\bar{x}_{i....}$ or the sums about depth $c \cdot \bar{x}_{i....}$ between "beech" and "spruce".

Significance of B (A): significant differences in the mean values $\bar{x}_{ij...}$ or the sums $c \cdot \bar{x}_{ij...}$ between plots within "beech" and/or "spruce".

Significance of C: significant differences in the mean values $\bar{x}_{..k..}$ between different depth intervals; that is, significant depth functions exist.

Significance of $A \times C$: significant differences in the mean values $\bar{x}_{i\ ..\ l}$, that is, in the depth functions between "beech" and "spruce". For some of the variables these mean values are shown in separate graphs for "beech" and "spruce", the asterisks indicating significant differences between "beech" and "spruce" according to the method of partitioning.

Significance of $B(A) \times C$: significant differences in the depth functions between the plots within "beech" and/or "spruce".

Significance of effect $A$ is diminished by greater variations in $B(A)$ and by significance of $A \times C$. In the first case it is necessary to discuss possible reasons for the variations of plots within trees in order to decide whether the experimental layout is unsatisfactory or, on the other hand, to detect factors of greater importance than the tree species for the variable under question. This discussion can only be of a hypothetical nature and will therefore be kept as short as possible.

## IV. Analytical Errors and Sampling Effects for Bioelement Content in Mineral Soil

From the bioelement content in mineral soil the exchangeable cations, $CEC_e$ and $N_t$ are shown in Table 3, expressed as mean values $\bar{x}$.....; the table contains further the sampling effects "sampling within plots within trees" $[P(A)(B)]$, the depth function of this sampling effect $[P(A)(B) \times C]$, and the analytical error, each calculated from the components of variance and given as a percentage of the mean value.

Table 3. *Bioelement content in mineral soil: mean values, sampling effects and analytical error*

| | H | Na | K | Ca | Mg | Al | Fe | Mn | CEC$_e$ | N |
|---|---|---|---|---|---|---|---|---|---|---|
| | | | | | $\bar{x}$ ... | | | | | |
| | | | | | $\mu$val/g | | | | | % |
| | 4.11 | 1.01 | 1.68 | 2.04 | 0.49 | 47.15 | 0.57 | 1,09 | 58.13 | 0.115 |
| | components of variance, expressed as deviations in % of $\bar{x}$ ... | | | | | | | | | |
| P(A)(B) | 27.2 | 56.7 | 19.3 | 16.3 | 25.1 | 6.1 | 25.8 | 7.9 | 7.4 | 8.2 |
| P(A)(B) × C | 36.5 | 41.3 | 16.2 | 13.9 | 15.8 | 12.5 | 46.1 | 32.5 | 11.0 | 11.5 |
| error | 10.2 | 19.5 | 8.5 | 10.6 | 24.8 | 9.0 | 28.0 | 8.7 | 7.8 | 0.7 |

For all exchangeable cations except Al the contents are in a very low range; this applies especially to Ca and is characteristic of strongly acid forest soils in Central Europe. The percentage analytical error, varying between 8.5 and 28% of the mean, is relatively large due to the low contents. Nevertheless, in all cases the two sampling effects are highly significant (*** with one exception: $P(A)(B) \times C$ for Mg: *). This means that the analytical error plays no role since, according to the statistical model, the main effects and interactions are tested against the sampling effects, this being the statistical consequence of the great spatial variability within plots. Of the two sampling effects, $P(A)(B)$ may be considered as representing the spatial variability of mineral soil stores, where the soil is not differentiated according to depth. If the various depths of mineral soil are considered, the spatial variability increases

considerably due to differences in depth functions within a site and to errors in sampling different depth intervals (cf. Table 3). BALL and WILLIAMS (1968), sampling the
0—15 cm horizon of uncultivated, unfertilized brown earths at two grassland sites
in circles of only 1 m radius on the same occasion, found coefficients of variation for
exchangeable cations averaging 33%. The comparison shows that the sampling
effects measured are not inherent to the sampling procedure, but reflect a spatial
variability of pedological and ecological significance.

Significant effects between beech and spruce (effects $A$ and $A \times C$) exist only for
H and Mn and to a lesser degree for Na (Fig. 1). The higher H content in the upper

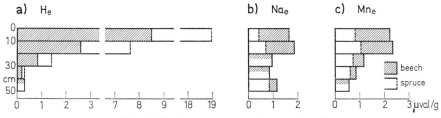

Fig. 1. Bioelement content in mineral soil

soil layers under spruce is the consequence of a higher H⁺ input into mineral soil, as
will be discussed in a later paper concerning bioelement cycling. The lower content
of Na and Mn in the same soil layers under spruce may be a consequence of the higher
acidity or of differences in cycling of these cations.

In the following statistical computation of the data, the analytical error has been
pooled with the sampling effects by feeding the mean values $\bar{x}_{ijkl}$ into the analysis
of variance. The statistical model is thereby reduced to

$$Y_{ijkl} = \mu + \alpha_i + \beta_{ij} + \gamma_k + (\alpha\gamma)_{ik} + (\beta\gamma)_{ijk} + \varepsilon_{ijkl}$$

## V. Carbon and Nitrogen Stores and pH

The depth functions of the variables are compiled in Fig. 2a—d; the significance
of differences between beech and spruce sites for each depth interval according to the
method of partitioning degrees of freedom and sums of squares is indicated by asterisks
(cf. III). The tree species significantly influences the depth functions: spruce having
higher C stores in the upper soil layers, especially in the organic top layer (O horizon),
but higher N stores only in the last one, wider C/N ratios through the whole depth
and lower pH down to 20 cm. Per ha and 50 cm, the C store amounts to 94 t for
beech and 131 t for spruce, whereas the N store is quite similar (5755 and 6067 kg
respectively). Consequently there are differences in the C/N ratio, indicating poorer N
nutrition at spruce sites. Assuming steady state, the ratio of carbon stores in the O horizon under spruce and beech (46 and 19 t C/ha respectively) allows the estimate
that the percentage of O horizon material, which is annually decomposed, is 2.4 times
larger under beech than under spruce. Since the higher C store down to 20 cm under
12*

spruce may be partly due to passive transport of water-soluble decomposition products from the O horizon into mineral soil, the factor of 2.4 will be too low. This indicates the different efficiency of litter-decomposing soil organisms at beech and spruce sites.

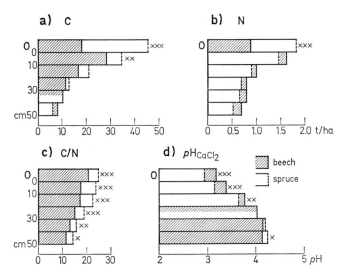

Fig. 2. Carbon and nitrogen stores and pH

## VI. Phosphate Stores

Like some of the already mentioned variables (especially $CEC_t$, carbon store and C/N ratio) the phosphate stores and C/$P_{org}$ ratio show highly significant effects for $B(A)$ and in case of $P_t$ and $P_{Al}$ for $B(A) \times C$. This indicates a large variability between the plots within one or both of the tree species. For a variable like total P store in soil, this variability cannot be caused by the variation in the age of the stands. If the parent material of the soil is the same, the only possibility for the variability in $P_t$ store is a P export from the site, maybe by soil erosion or by human misuse. As pointed out elsewhere in greater detail (Ulrich et al., 1969b), various of the effects mentioned may apply simultaneously. The fact as such is of ecological significance, as may be demonstrated for the C/$P_{org}$ ratio. This ratio can be used as an indicator for the amount of P mineralized by decomposing a distinct amount of organic matter: the closer C/$P_{org}$ the higher should be the P mineralization and, therefore, the contribution of organic bound phosphorus to the nutrition of plants. The analysis of variance shows that site-effects, as mentioned before, are of greater importance to P metabolism than actual tree species.

$P_{Al}$ (Table 4) constitutes only 6% of $P_t$ and shows a distinct depth function (increase with depth) in the mineral soil. This finding is confirmed by the solubility product of Al phosphates (Table 4), calculated from the data for equilibrium soil solution according to the equation

$$pK = (pH + pH_2PO_4) - 3 (pH - 1/3 \, pAl)$$

Table 4. *Aluminium phosphate stores* ($P_{Al}$) *and solubility product of soil aluminium phosphates* (pK)

| Depth cm | $P_{Al}$ kg/ha | | | pK | | |
|---|---|---|---|---|---|---|
| | beech | spruce | F-test | beech | spruce | F-test |
| 0—10 | 16.7 | 17.0 | n. s. | 3.09 | 3.11 | n. s. |
| 10—20 | 7.9 | 6.6 | n. s. | 2.34 | 3.05 | *** |
| 20—30 | 11.3 | 20.9 | *** | 2.47 | 2.44 | n. s. |
| 30—40 | 22.0 | 37.5 | *** | 2.83 | 2.33 | * |
| 40—50 | 24.1 | 34.8 | *** | 2.76 | 2.57 | n. s. |

Crystalline $AlPO_4$ (variscite) has a pK-value of 2.5 (LINDSAY et al., 1959) and amorphous $AlPO_4$ of 0.5 (TAYLOR et al., 1962). By comparison with these values, Table 4 shows that at 0—10 cm for beech, 0—20 cm for spruce the solubility product of variscite is exceeded, indicating that variscite is not a stable solid phase. In deeper layers of spruce soil, pK is close to 2.5; Al phosphatase should therefore be more stable there and this is confirmed by the higher $P_{Al}$ stores. The difference between beech and spruce may in this case be a function of root density since the Al phosphates can be expected to be the most readily available mineral phosphate pool in these acid soils. If this is true, it follows that the P nutrition of spruce from mineral phosphate sources is poorer than that of beech, this being a consequence of different root density in deeper soil layers.

## VII. Mobilizable Cation Stores

The amounts of mobilizable cations including the sum exchangeable cations, $CEC_e$, show in all cases significant depth functions, either for effect $C$ or effect $A \times C$. The comparison between beech and spruce soils in Fig. 3 is therefore done for the different depth intervals, whereas the first depth interval corresponds to the O horizon.

In most cases (exceptions: Na, Mn) the mobilizable cation stores reach their highest values in the O horizon with a marked difference between beech and spruce due to the larger amounts of O horizon material under spruce. Table 5 shows the bioelement concentration in the O horizon, expressed in g-atoms/t C. For any element the concentration is higher by a factor of 1.25 (Si) to 1.9 (Mg) under beech than under spruce, indicating a higher retention power of beech humus for the bioelements passing the soil surface by litter fall or crown washout. For the same reason, the cation concentrations increase in the order monovalent-divalent-trivalent with the exception of Mn, which participates in the cycling processes in beech stands to a small

Table 5. *Bioelement concentration in O-horizon material*

| | Na | K | Ca | Mg | Al | Fe | Mn | N | P | Si |
|---|---|---|---|---|---|---|---|---|---|---|
| | | | | | g-atom/t C | | | | | |
| beech | 35.9 | 131 | 130 | 81 | 787 | 456 | 19.6 | 3401 | 107 | 23300 |
| spruce | 23.9 | 75 | 75 | 43 | 524 | 256 | 13.4 | 2929 | 76 | 18700 |
| F-test A | * | * | * | ** | n. s. | * | n. s. | n. s. | n. s. | n. s. |
| B(A) | n. s. | n. s. | ** | n. s. | n. s. | n. s. | *** | n. s. | *** | n. s. |

degree only. Especially for Mg and Ca, the accumulation of mobilizable stores in the O horizon causes an unbalanced depth distribution which is characterized by what may be called noise heaviness. This is especially true for Mg and is indicative of some kind of stress situation, since the mineral soil seems to be more or less exhausted in respect to this element. The noise heaviness in depth distribution is in all cases more pronounced under spruce than under beech. The differences in stores in mineral soil under beech and spruce (cf. Fig. 3) are the immediate consequence of the greater accumulation in O horizon under spruce. Except for H, $CEC_e$ and Mn (cf. IV), the sum of stores about all depths intervals are quite similar between spruce and beech (no significance of effect $A$).

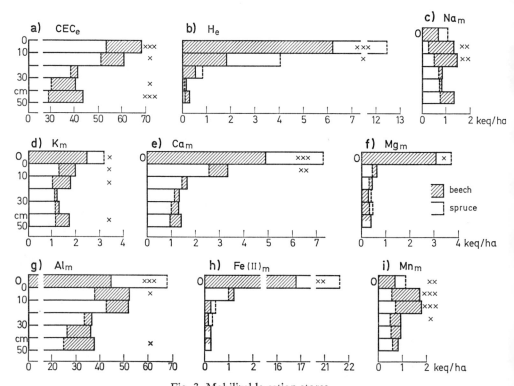

Fig. 3. Mobilizable cation stores

# VIII. Equivalent Fractions of Exchangeable Cations

The possibility of expressing exchangeable cation contents or stores as dimensionless ratios facilitates comparisons between different soils, because the ratios are independent of $CEC$ and therefore variations in $CEC$ due to different organic matter content or soil texture no longer play a role. The store expressed, e.g. in keq/ha, clearly represents a capacity or, in other words, an extensive soil property. The equivalent fractions of exchangeable cations $X^S$, on the other hand, represent an intensive soil property which reflects only the soil state. Since the soils selected for the study show no significanct difference in $CEC_t$ between beech and spruce, no gain of information is to be expected from the use of equivalent fractions.

The overall picture (Fig. 4) is that common to strongly acid forest soils in Central Europe; in the soil layers of lower humus content Al dominates with 80—90%, whereas H is present in greater fractions only in the upper soil layers with higher humus content. K and Ca amount to 3—4%, Mg to 1% of $CEC_e$.

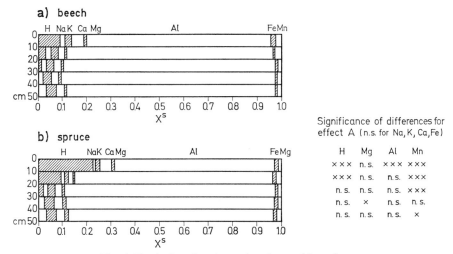

Fig. 4. Equivalent fractions of exchangeable cations

## IX. Cation Equivalent Fractions in Equilibrium Soil Solution

Due to a large error which, according to the statistical model, includes the sampling effects, the ion concentrations in ESS show no significance for the differences between beech and spruce. These values will therefore not be recorded, and the survey will be given by the cation equivalent fractions in ESS. As can be seen from Fig. 5,

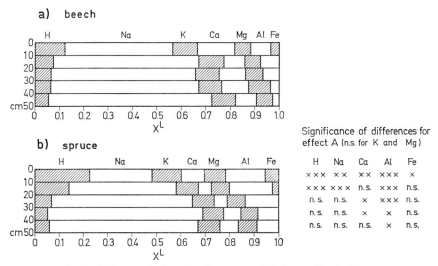

Fig. 5. Cation equivalent fractions in equilibrium soil solution

Na is the dominant cation in the soil solution with a significantly greater fraction in beech than in spruce soils, especially in the upper layers of the mineral soil. This corresponds to the higher contents and stores of exchangeable Na in the same layers under beech. The high sodium percentage in ESS, which has been demonstrated also by le Roux et al. (1967), cannot be explained by soil properties, but seems to be caused by discrimination of Na during ion uptake by roots (Ulrich, 1966b). This problem deserves further investigation.

Further significant differences between soils under beech and spruce exist for H and Al, both cations indicating soil acidity and being higher under spruce. Whereas the higher H percentage corresponds to content or store of exchangeable H, the Al percentage does not. This indicates differences in the binding strength of exchangeable Al which will be discussed later.

## X. Schofield's Potentials in Equilibrium Soil Solution

If a soil contains $n$ exchangeable cations, which are determined in ESS, a series of $n$ $(n-1)$ Schofield's potentials $(SP)$ can be calculated. For complete description of ESS a set of $n$-1 $SPs$ is enough, if correctly chosen, because this is the least number of $SPs$ necessary to calculate all the others.

Since in the ESS the cations H, Na, K, Ca, Mg, Al and Fe have been determined, the total series corresponds to 42 $SPs$, the means $x$ . . . . of which are compiled in Table 6 in matrix form. As can be seen, the $SPs$ on the left of the diagonal have the same value as the corresponding $SPs$ on the right of the diagonal, but with opposite sign. Only the $SPs$ right of the diagonal will therefore be discussed further.

As a result of the analysis of variance, the error of $SPs$, which includes the sampling effects or the spatial variability of the plots, lies between 0.05 and 0.29, i.e. is of the same order of magnitude as the errors in pH (0.08, Table 3). The errors exceeding 0.2 belong to $SPs$ including pNa or pK.

The $SPs$ have so far been little used in ecological soil investigations. Their informational value follows from their definition, which has been given above. Positive values indicate that the reduced activity ratio $R$ of the cation pair is less than 1. For the lime potential pH-1/2pCa, the values determined in soils may vary between $>6$ and $<1$ ($a_H/a_{Ca} < 10^{-6}$ or $> 10^{-1}$), whereas the pH value varies between $>7$ and $<3$ ($a_H < 10^{-7}$ or $> 10^{-3}$). The significant difference between beech and spruce soils in respect to pH $-1/2$ pCa (1.87 vs. 1.62) means therefore, that the ratio $a_H/\sqrt{a_{Ca}}$ is significantly narrower under spruce than under beech. Whether this has any importance for plants growing on the respective soils is an open question.

Significant effects for the difference between beech and spruce soils $(A, A \times C)$ exist for pH-pNa, pH-1/2 pCa, pH-1/2 pMg, pH-1/3 pAl, pNa-pK, pNa-1/2 pMg, pNa-1/2 pFe, 1/2 pCa-1/3 pAl, 1/2 pCa-1/2 pFe, 1/2 pMg-1/3 pAl, 1/2 pMg-1/2 pFe and 1/3 pAl-1/2 pFe; that is for 24 out of 42 $SPs$. If this is compared with the number of significant effects found for ion concentration in ESS (0 out of 7) and cation equivalent fractions in ESS (3 out of 7), the suitability of $SPs$ for such a comparison becomes obvious.

The differences between soils under beech and spruce can be grouped as follows ($j =$ any cation): beech $>$ spruce for pH-$j$, 1/3 pAl-$j$, 1/2 pFe-$j$ and beech $<$ spruce for pNa-$j$, pK-$j$, 1/2 pCa-$j$, 1/2 pMg-$j$. There are necessarily some exceptions but the groups represents the general tendency. This means that the reduced ratios in the case of the cations responsible for soil acidity in the numerator of the ratio are narrower in the soils under spruce than under beech, whereas the opposite is true of the so-called basic cations in the numerator. This is a consequence of the higher soil acidity under spruce, and the *SPs* show that the chemical potential of nearly all cations in soil solution is influenced by the higher soil acidity.

Table 6. *Matrix of Schofield's potentials (mean values × . . . .). The asterisks indicate the significance of effect A (1st row) or A × C (2nd row)*

| | — pH | — p Na | — p K | $-\frac{1}{2}$pCa | $-\frac{1}{2}$pMg | $-\frac{1}{3}$p Al | $-\frac{1}{2}$pFe |
|---|---|---|---|---|---|---|---|
| pH — | | 0.83 n.s. * | 0.12 | 1.75 ** *** | 1.74 * * | 2.35 n.s. ** | 1.13 |
| p Na — | —0.83 | | —0.73 n.s. * | 0.91 | 0.89 n.s. *** | 1.51 | 0.29 n.s. * |
| p K — | —0.12 | 0.73 | | 1.03 | 1.63 | 2.24 | 1.02 |
| $\frac{1}{2}$pCa — | —1.75 | —0.91 | —1.03 | | —0.01 | 0.60 *** n.s. | —0.62 * n.s. |
| $\frac{1}{2}$pMg — | —1.74 | —0.89 | —1.63 | 0.01 | | 0.61 *** n.s. | —0.61 |
| $\frac{1}{3}$pAl — | —2.35 | —1.51 | —2.24 | —0.60 | —0.61 | | —1.22 |
| $\frac{1}{2}$pFe — | —1.13 | —0.29 | —1.02 | 0.62 | 0.61 | 1.22 | |

There are significant differences between depths for all *SPs* except pK-1/2 pMg and pK-1/3 pAl, and for the *SPs* pH-$j$, pNa-pK and pNa-1/2 pMg there are significantly different depth functions between beech and spruce soils. The depth functions are not shown in this paper, they can be allocated to the same groups- pH-$j$, 1/3 pAl-$j$ and 1/2 pFe-$j$ show, like the depth function of pH, an increase with increasing depth, especially from the first to the second or third layer, whereas the other groups of *SPs* decrease more or less steadily with depth. Some exceptions must naturally occur if cations of the same group are combined in one *SP*. The differences between beech and spruce are restricted to the upper 20 cm of the soil, with the exception of *SPs* including Ca or Mg, which may be significantly different over the whole depth.

The question remains what information can be deduced from the *SPs*. For a few *SPs* some information of an ecological nature can be found in the literature, dealing mainly with relationships between *SP* and yield and/or nutrient content of agricultural crops. But the results are contradictory, which is quite understandable, since a *SP* is comparable to a pressure or a tension. This means that, for investigations of the kind mentioned, either the *SP* must be constant over the whole growth period, or the mass of nutrient present in the system with a given *SP* must be considered and not the *SP* itself. But as an example of the possible ecological information value of *SPs* the statement of WOODRUFF (1955) may be cited: that for agricultural crops values of pK-1/2 pCa $> 2.6$ indicate potassium deficiency, values between 1.8 and 2.2 optimum potassium nutrition, and values $< 1.5$ either potassium excess or calcium deficiency. The soils of this investigation belong to the last group.

Another kind of information the *SPs* can yield regards the influence of soil solution moving down the soil profile on the composition of exchangeable cations. Since the total of *SPs* characterizes the composition of the soil solution, being in equilibrium with a distinct composition of exchangeable cations, a depth function of *SPs* necessarily means that transport of the soil solution through the solum will be accompanied by cation exchange reactions. If I and II indicate different positions of ESS in soil, and the ESS is transported from position I to position II without intermingling, the chemical potential $\Delta \mu\, A/B$ for the exchange reaction including the cations $A$ and $B$ and caused by this transport is (ULRICH, 1969a)

$$\Delta \mu_{A/B} = -1.364\,(SP^{I} - SP^{II})\ \text{kcal}.$$

This accounts for all *SPs*, so that $n(n-1)\Delta \mu$'s have to be considered. The chemical potential influencing the ion $i$ is equal to

$$\Delta \mu_i = \sum_{j=i}^{n} \Delta \mu_{i/j}$$

where $j$ symbolizes the other cations present. $j$ runs from 1 to $n$.

Table 7. *Chemical potential for cation exchange reaction by vertical downward transport of equilibrium soil solution (mean values $\bar{x}_{i\,.\,k}.$ resp. $\bar{x}\,.\,.\,_k.$ $\Delta \mu$ kcal/mole. B = beech, S = spruce*

| ± cm | H B | H S | Na B | Na S | K B, S | Ca B | Ca S | Mg B, S | Al B, S |
|---|---|---|---|---|---|---|---|---|---|
| $\dfrac{0-10}{10-20}$ | $+1.99$ * | $+1.26$ | $-1.85$ | | $-0.05$ | $+0.82$ | | $-0.18$ | $-0.37$ |
| $\dfrac{10-20}{20-30}$ | $+0.62$ ** | $+2.19$ * | $+0.35$ * | $-1.61$ | $-0.44$ | $-0.11$ | | $-0.37$ | $+0.14$ |
| $\dfrac{20-30}{30-40}$ | $+0.08$ | $+1.06$ | $-0.20$ | | $-0.01$ | $-0.29$ | | $-0.30$ | $+0.23$ |
| $\dfrac{30-40}{40-50}$ | | $-0.01$ | $-0.35$ | | $-0.95$ | $+0.54$ ** | $-0.37$ | $+0.16$ | $+0.20$ |
| A × C | * | | n. s. | | * | * | | n. s. | n. s. |

The results of this calculation are shown in Table 7 for the means $\bar{x}_i \ldots$ or $\bar{x} \ldots$ , omitting Fe due to its low fraction in exchangeable cations and soil solution. Since the deviation in $SPs$ due to effect $C$ and $A \times C$ is of the same size as the error, the differences, if there are any, between $\Delta \mu$'s are of low significance, the effects $A, B(A)$, $C$ and $B(A) \times C$ being not significant throughout. In the table positive values for any cation indicate a driving force for being sorbed by exchange of any other cation, preferably of the cations showing a high negative $\Delta \mu$. Below 2.2 kcal/mole all $\Delta \mu$'s are quite low, being highest in the top soil and decreasing with increasing depth. It may be concluded from the table that there is in the top soil a weak tendency for the exchange of H and Na ions, H being sorbed and Na being desorbed. For spruce this tendency is visible down to 30 cm depth, for beech only to 20 cm.

## XI. Gapon Coefficients

The Gapon equation, as mentioned before, describes cation exchange equilibria of soils within the limits of constant exchanger properties. If the properties of cation exchangers in soils vary with depth, or depend upon the composition of exchangeable cations, the Gapon coefficient will not be a constant but a function of parameters expressing exchanger properties.

The numerical value of the Gapon coefficient is a measure of the selectivity of cation binding at exchanger surfaces. If the ratio $A_s/B_s$, where $A_s$ and $B_s$ are expressed in equivalents per unit weight of soil, is equal to the reduced ratio $R_{A/B}$, the Gapon coefficient is 1, indicating that both cations are bound without any preference or discrimination. For Gapon coefficients $k^G_{A/B} > 1$, cation $A$ is preferred to $B$ by the exchanger, the selectivity for $A$ being the greater the higher the numerical value of $k^G_{A/B}$. For $k^G_{A/B} < 1$, cation $A$ is discriminated against $B$ by the exchanger, the selectivity for cation $B$ being the higher the lower the numerical value of $k^G_{A/B}$.

In Table 8 the mean values $x \ldots$ of the Gapon coefficients are listed in the form of a matrix. Note that the values on the left of the matrix are the reciprocal of the values on the right, so that only the latter need be considered. $k^G_{Na/K}$ allows an estimation of the fraction of exchangeable K bound, like Na, at organic exchangers or at planar sites of clay minerals (ULRICH, 1966a). From the values listed in Table 8 it follows that only 9% of exchangeable K is bound at planar sites, the rest being assumed to be bound at the opening edges of the illite sheets. The hig hvalues of $k^G_{K/j}$ (second row of the matrix, $j =$ any other cation) show that the specificity of these sites is valid for all cations including Al. It can also be seen that Ca can compete more easily with K than Mg.

A fact which deserves special attention is the high selectivity of Na in comparison to Ca and Mg. This uncommon exchange equilibrium of Na requires further investigation. It may be that the Gapon coefficients do not reflect an equilibrium so much as a steady state, caused by a continuous input of sodium ions into the soil solution. The source of this input can only be the root discriminating against Na during ion uptake. This hypothesis would also explain the high equivalent fraction of Na in soil solution (cf. ULRICH, 1966b).

According to its valency, Al is selectively bound in respect to Na, Ca and Mg. Ca is selectively bound in respect to Mg, but discriminated against relative to all other cations.

In case of Ca/Al and Mg/Al, the selectivity of Al is lowered in spruce soils, which is in accordance with the significantly higher equivalent fractions of Al in soil solution. In the upper 20 cm the lower binding strength of Al under spruce may be caused by the higher humus content and the lower pH, in the deeper soil layers by the differences in soil texture mentioned and possibly clay composition.

Table 8. *Matrix of Gapon coefficients (mean values $\bar{x}$ ....).*
*The asterisks indicate the significance of effect $A$ (1st row) or $A \times C$ (2nd row)*

|      | /Na  | /K    | /Ca  | /Mg  | /Al          |
|------|------|-------|------|------|--------------|
| Na/  |      | 0.09  | 33   | 13   | 0.53         |
| K/   | 11   |       | 43   | 178  | 6.6          |
| Ca/  | 0.3  | 0.02  |      | 4.2  | 0.17 *** *** |
| Mg/  | 0.08 | 0.006 | 0.24 |      | 0.04 *** n. s. |
| Al/  | 1.9  | 0.15  | 5.7  | 22   |              |

# XII. Discussion

One important feature of the analysis of variance of the data is the significance often found for the effect "plots within tree species" $[B(A)]$ and the interaction of this effect with the depth $[B(A) \times C]$. As mentioned in sections V and VI, the variation of plots within tree species can be traced back to variations in the parent material of the soils and to differences in earlier land use. Effects of this kind can overshadow the influence of tree species on soil. Since the soils for this study have been selected according to parent material and soil type, existing variations in parent material are restricted mainly to the subsoil, and variations in the top soil are due to earlier land use. According to the statistical analysis, this effect is at least of equal if not greater importance for present soil properties than the actual tree species.

One of the basic differences between soils under beech and spruce is connected with soil acidity. All variables which are a measure of soil acidity [pH, content and store of exchangeable H, equivalent fraction of exchangeable H and of H and Al ions in equilibrium soil solution and the relevant Schofield's potentials pH-$j$, 1/3 pAl-$j$ and 1/2 pFe-$j$ ($j$ = any other basic cation)] indicate a significantly higher soil acidity in the upper 20 cm of mineral soil under spruce. The tendency towards further soil acidification exists under beech as well as spruce, but extends more deeply into the mineral soil under spruce.

The higher soil acidity under spruce influences the Schofield's potentials of all cations in soil solution. This should have consequences for bioelement uptake by the stands but, for lack of relevant research, it is not yet possible to make statements about the kind and extent of these influences. It is hoped that the investigations into bioelement cycling and the bioelement inventory carried out as part of the Solling Project will give some results in this respect.

The other basic difference between soils under beech and spruce is caused by the properties of spruce litter and influences the depth functions of all bioelement stores in soil. Due to the slower decomposition of litter, the organic matter store in the O horizon under spruce is more than twice as great as under beech. Greater amounts of the bioelements reaching the soil surface within the cycling process are therefore retained in the O horizon under spruce, in spite of the higher retention power of beech humus. The amounts accumulated in the O horizon are lacking in the mineral soil, so that for most of the bioelement investigated the stores under spruce are greater in the O horizon, but lower in the mineral soil. The consequence for N is a higher C/N ratio throughout the whole solum under spruce, indicating worse N nutrition conditions. For P, the same is true only for the O horizon, similar effects in deeper soil layers being suppressed completely by the wide variations of plots within tree species. Only the solubility of Al phosphates in the upper soil layers indicates that the P nutrition of spruce from mineral phosphate sources is poorer than that of beech. Overall, the higher accumulation of bioelement stores in the O horizon under spruce may affect the root distribution.

These results agree in general with earlier work (GENSSLER, 1959; EVERS, 1969). The more sophisticated methods used in this investigation resulted in the detection of some sofar unknown effects, but only in certain cases was it possible to explain these effects or to draw an ecologically useful conclusion. This simply indicates that methods developed to characterize the soil as a natural body do not necessarily yield information of ecological significance. There is a great lack of meaningful ecological methods of characterizing the bioelement state of a site. It can be expected that investigations into bioelement cycling will suggest other ways which may be followed in developing such methods, since the main interest for ecology lies in what may be described as the fate of bioelements at a site. Methods available in the field of soil chemistry will be indispensable for developing mathematical models of the fate of bioelements, but they are not appropriate for characterizing it.

# References

ATANASIU, N., BEUTELSPACHER: Bohrgerät für die kontinuierliche und volumetrische Boden-entnahme unter natürlichen Bedingungen. Z. Acker- u. Pflanzenbau **113**, 338—341 (1961).

BALL, D. F., WILLIAMS, W. M.: Variability of soil chemical properties in two uncultivated brown earths. J. Soil. Sci. **19**, 379—399 (1968).

CHANG, S. C., JACKSON, M. L.: Fractionation of soil phosphorus. Soil Sci. **84**, 133—144 (1957).

EVERS, F.-H.: Untersuchungen über die Auswirkungen des Fichtenreinanbaus auf Para-braunerden und Pseudogleye des Neckarlandes. IV. Chemische Untersuchung ober-flächennaher Bodenbereiche. Mitt. Verein Forstl. Standortskunde u. Forstpflanzenzüch-tung **19**, 90—92 (1969).

Genssler, H.: Veränderungen von Boden und Vegetation nach generationsweisem Fichten-anbau. Diss. Univ. Göttingen 1959.

Khanna, P. K., Ulrich, B.: Phosphatfraktionierung im Boden und isotopisch austausch-bares Phosphat verschiedener Phosphatfraktionen. Z. Pflanzenernähr., Düngung u. Boden-kunde. 117, 53—65 (1967).

le Roux, I., Summer, M. E.: Studies on the soil solution of various Natal soils. Geoderma 1, 125—130 (1967).

Lindsay, W. L., Peech, M., Clark, J. S.: Solubility criteria for the existence of variscite in soils. Soil Sci. Soc. Am. Proc. 23, 357—360 (1959).

Moss, P.: Some aspects of the cation status of soil moisture. I. The Ratio Law and soil moisture content. Plant and Soil 28, 99—113 (1963).

Richards, L. A. (ed.): Diagnosis and improvement of saline alkali soils U.S. Dept. Agric. Handbook No. 60, 1954.

Riehm, H., Ulrich, B., Ulrich, M.: Schnelle Bestimmung der Kationensorptionskapazität. Landw. Forschung 6, 95—105 (1954).

Steel, R. G. D., Torrie, J. K.: Principles and procedures of statistics. New York: Mc Graw-Hill 1960.

Taylor, A. W., Gurney, E. L.: Solubility of amorphous aluminium phosphate. Soil Sci. 93, 241—245 (1962).

Ulrich, B.: Kationenaustausch-Gleichgewichte in Böden. Z. Pflanzenernähr., Düngung u. Bodenkunde 113, 141—159 (1966a).

— Selectivity and discrimination in ion uptake under field conditions. Int. Atomic Energy Agency Tech. Rep. Ser. 65, 121—127 (1966b).

— Chemische Potentiale beim Transport von Bodenlösung durch das Solum. Z. Pflanzen-ernährung u. Bodenkunde 123, 181—186 (1969a).

— Khanna, P. K.: Ökologisch bedingte Phosphatumlagerung und Phosphatformenwandel bei der Pedogenese. Flora Abt. B 158, 594—602 (1969b).

— — Methodische Untersuchungen über Kationengehalt der Bodenlösung und Schofield-sche Potentiale. Göttinger Bodenkdl. Ber., in press 1970.

Woodruff, C. M.: Ionic equilibria between clay and dilute salt solutions. Soil Sci. Soc. Amer. Proc. 19, 36—40 (1955).

# X. Investigations of the Content and the Production of Mineral Nitrogen in Soils

M. RUNGE

## I. Introduction

Nitrogen is taken up by higher plants mainly in the form of ammonium and nitrate ions. The nitrogen supply to the vegetation in natural or near-natural sites therefore depends on the formation of ammonium in the course of the decay of organic nitrogen compounds in litter and humus. The additional contribution of ammonium and nitrate from other sources to the natural N cycle of the sites, for instance with the rainfall, is generally of minor importance.

In many soils ammonification is followed by nitrification, i.e. oxidation of the primarily produced ammonium to nitrate. This process is important for higher plants because nitrogen in this case is available not only in the form of the cation, but in the form of the anion, too. Under certain conditions, some species prefer nitrate or a mixed nitrate-ammonium nutrition (EVERS, 1964; BOGNER, 1968).

Part of the mineral nitrogen produced (Nmin), preferably ammonium, is consumed by the soil microorganisms (JANSSON, 1958). As they are more successful than the higher plants in competing for Nmin (BARTHOLOMEW and CLARK, 1950; JANSSON, 1958; ZÖTTL, 1960a), of the total Nmin production (gross mineralization) the higher plants can use only that part (net mineralization) which exceeds the microbial demand.

The natural Nmin content of the soil (field Nmin content) represents the difference between net mineralization and Nmin uptake by the vegetation. Consequently it does not permit conclusions on the Nmin supply to the vegetation. For that purpose the net mineralization itself has to be determined.

Because of methodological difficulties, this important production factor has been little investigated up till now. With indirect methods, for instance, using the C/N ratio of the humus, conclusions can often be drawn on the N supply to the vegetation. But these methods are restricted to certain site conditions. Moreover, they are neither suitable for analyses of the causal relations between the N supply and other site factors nor for investigation of essential details of the nitrogen supply in different sites.

The investigations in the Solling Project have two main objects: 1. to examine critically methods used for the determination of net mineralization as regards the quantitative interpretation and reliability of the results; 2. to investigate differences in the nitrogen supply of the different sites, i.e. the yearly Nmin supply, the progress of net mineralization during the year and its vertical distribution in the soil profile, the $NH_4/NO_3$ ratio and its relations to other site factors.

In this first publication mainly the methods will be described. Some results from the most intensively examined beech stand will illustrate their application.

# II. Methods

## 1. Determination of Net Mineralization

Hesselman (1917) was the first to attempt the comparison of mineral nitrogen production in different forest soils. Keeping soil samples under certain conditions in the laboratory, he took the increase of mineral nitrogen per time unit as a measure of net mineralization capacity of soils. This "incubation method" makes it possible to carry out comparative research at several sites in a relatively short time (Zöttl, 1958, 1960b; Runge, 1965). Ehrhardt (1959) and Ellenberg (1964) introduced modifications of this method: they stored the samples at the site so as to observe the special site conditions as far as possible.

In this investigation the principle of the method introduced by Ellenberg (1964) is used. Soil samples are taken periodically throughout the year and stored at the site for 6 weeks. The samples are kept in polyethylene bags to prevent penetration of roots and leaching of mineral nitrogen. The difference between the content of mineral nitrogen at the beginning and at the end of these 6 weeks gives a measure of net mineralization.

## 2. Treatment of Samples

Net mineralization in the soil was investigated first to a depth of 20 cm and later 40 cm of the mineral soil. Sampling was carried out at several depths, each depth being handled separately (Fig. 1a—c). In the organic layer, 4 horizons were distinguished and designated $L$, $F_1$, $F_2$ and $H^*$. On every sampling date 20 single samples, each from an area of about $20 \times 20$ cm, were taken and mixed by hand in a plastic bowl. Roots were removed as far as possible.

Similarly, 20 single samples (about 100 cm³) were taken from the uppermost part of the mineral soil (0—6 cm). To obtain samples from deeper layers, usually 3 pits (about $60 \times 40$ cm wide and 34 cm deep) were dug. Material was taken both from the walls and the bottom of these pits, and each sampling depth was mixed separately.

One part of each mixed sample was brought to the laboratory and the mineral nitrogen and the water content was determined immediately. Another part was placed in a polyethylene bag ($20 \times 32$ cm, 50 $\mu$) to about 1/3 of the bag's capacity. The bags were knotted tightly, to leave as little air inside as possible, and subsequently replaced in the hollows of the pit walls from which the material had been taken. The pits were then filled up again. In the organic layer the bags were placed in the appropriate horizons, as flat as possible, and, including the material from the $L$ horizon, covered with litter.

The polyethylene bags prevent removal of mineral nitrogen during storage both by the roots of higher plants and by leaching. As the foil is impermeable to water and nearly impermeable to water vapour, the water content remains constant, whereas in open vessels uncontrollable changes could occur. On the other hand, the permeability to $O_2$ and $CO_2$ is sufficient, and no impairment of the net mineralization (Runge, 1970) nor of the specially $O_2$-dependent nitrification (Eno, 1960) could be ascertained.

After in general 6 weeks' storage of the bags, the mineral nitrogen content of the samples was analysed and the water content determined again, as the bags had some-

---

* For the beech site, they correspond to the following horizons distinguished by Babel (see chapter Va): L = $L_n$, $F_1 = L_v + F_r$, $F_2 = F_m + H_r$, H = $A_{hh}$.

times been gnawed by soil fauna. Samples in heavily damaged bags and those with a substantially changed water content were discarded. Generally 3 bags per horizon were stored, to lower the risk of losing data due to destruction of the bags.

## 3. Analysis

For the analysis the soil was first passed through a 6 mm mesh sieve. 5 or 10 g of the sieved, natural moist material was weighed into a 300 ml Erlenmeyer flask and 40 or 80 ml of a 1% $KAl(SO_4)_2$ solution was added. After shaking for 1/2 hour, the suspensions were filtered with a Witts filter apparatus into 100 ml Erlenmeyer flasks. The filtrate was either used for analysis immediately or stored in a refrigerator at about 6° C.

The analysis of ammonium and nitrate in the filtrate has been carried out in two different ways:

At first colorimetric methods were used. Nitrate was determined with 2.4-xylenol (SCHARRER and SEIBEL, 1956; prescription RUNGE, 1964), and ammonium after CONWAY (1962). In the latter case Nessler's reagent was used at first, later on the indophenol reaction after Berthelot (FAWCETT and SCOTT, 1966), following a slightly modified prescription of YERLY (1970). This reaction is about 10 times more sensitive than the Nessler reaction, but in our experience, more reliable. The colour, for instance, remains stable for more than 24 hours.

In the course of the investigations it became necessary to increase the number of analyses per time unit. Otherwise, as will be shown later, some important questions of method would not be solvable.

The distillation method, described by BREMNER and EDWARDS (1965) and BREMNER and KEENEY (1966) proved to be suitable. When filtrates of soil suspensions were used, the results were the same as with the colorimetric methods. By using 4 distillation apparatuses, one together with an automatic titration apparatus (Combititrator 3D, Metrohm), the time needed to perform a certain number of analyses was reduced to less than $1/_3$ of that needed for colorimetric methods. The $NH_4$ and $NO_3$ content of about 50 filtrates per day can be determined by one person, including preparing and washing up glassware, but not setting up the filtrates.

At first we hoped to reduce the time still more by omitting filtration and using the soil suspension directly, or the decanted, supernatant liquid. But in both cases the results were higher than with filtrates, especially when using suspensions. BREMNER and KEENEY (1966) did not find differences, when comparing the supernatant liquid and the filtrate, but they used air-dried soil, while we must take natural moist soil. Probably, some destruction of organic N-compounds occurs during steam distillation, when fresh soil is used. Methodical investigations and the prescription of the procedure for using the distillation method for the analysis of natural moist soil will be described in a special publication (GERLACH, in preparation). As the investigations, described in the present publication, were done by colorimetric methods, the distillation method is not dealt with in detail.

## 4. Determination of Water Content and Volume-Weight, or "Area-Weight"

The water content was determined by drying a part of each sieved sample at 105° C in an oven.

To determine the volume-weight of the organic horizons, sheet metal frames (30 × 30 cm, 5 cm high) were inserted into the organic layer, and the horizons inside the frame were removed one after another, sieved, and dried as described above. Corresponding determinations in the mineral horizons were carried out by taking samples in 100 cm³ cylinders (25 cm², 4 cm high). Likewise, the material was sieved before drying. Roots, bigger twigs and stones were removed.

Thus the weight of the "sievable fraction" of the material was established. This fraction contains all the components that are mineralizable during the storage of the samples.

## 5. Calculation of Nmin Content

As natural moist soils with different and varying water contents are used for analysis, it is necessary to convert the results to a dry weight basis to make them comparable. It has also to be remembered that, although a constant amount of $KAl(SO_4)_2$ solution is added to the sample, it contains a varying quantity of soil water. This causes a dilution of the filtrate that cannot be neglected, especially in the organic horizons with water contents up to 400%.

For the conversion of the results, therefore, we need a formula that for a given water content of the moist soil sample takes into account the corresponding amount of dried soil as well as the dilution effect (see e.g. STÖCKER, 1968). When analysing a large series, as in the present investigation, it is better to make the calculation in several steps. For any given proportion between the amount of natural moist soil and solvent, correction factors can be calculated which eliminate the dilution effect of any given water content. The use of these tabulated correction factors significantly simplifies the calculations.

The calculations were made in the following order:

a) Conversion of the extinction values of the photometer in "preliminary $NH_4$ and $NO_3$ contents" of the natural moist soil by means of conversion factors, established previously with standard solutions.

b) Correction of the dilution effect by multiplying the preliminary Nmin content by the correction factor for the water content of the sample.

c) Conversion to terms of dry weight: (mg Nmin/g moist soil) × (% water content + 100) = mg Nmin/100 g dry soil.

These values were additionally converted to values based on the volume of soil with undisturbed structure or on the area of the stand. Because of the variation in volume-weight and depth of the horizons, a comparison of the Nmin production of horizons and sites is possible only on an area basis.

## III. Some Results

### 1. Nmin Content of the Soil under Field Conditions

#### a) Horizontal Distribution

When the "incubation method" is used to determine net mineralization capacity, the field Nmin content of the sample can be neglected in cases where it is small compared with the content after incubation (ZÖTTL, 1958; RUNGE, 1965). In samples stored at the site, the Nmin increase is significantly smaller than with laboratory incubation. The field Nmin content, compared with the increase at the site, is in most cases relatively high and has to be taken into consideration.

In the soils investigated in the Solling Project, the field Nmin content was very high, compared with the information in the literature and our own, earlier experience. Therefore its periodical determination was not only necessary for the investigation, but moreover of general importance.

First of all, the horizontal variations of the Nmin contents in the different layers had to be investigated. Table 1 shows the standard deviations ($s$) on the basis of 20

Table 1. *Mean values ($\bar{x}$) and standard deviations ($s$) of the field Nmin content (mg Nmin/100 g dried soil) of the organic soil layer of the beech stand. Sampling: 17. 7. 1967*

| Horizon | L | $F_1$ | $F_2$ | H |
|---|---|---|---|---|
| n | 20 | 20 | 20 | 20 |
| $\bar{x}$ | 9.1 | 16.8 | 9.7 | 2.5 |
| $s$ | 4.0 | 4.5 | 3.2 | 1.0 |
| $s\%$ | 44 | 27 | 33 | 42 |

single samples from each of the 4 horizons in the organic layer. Similar determinations made earlier in the same layers showed relative standard deviations ($s\%$) of the same order of magnitude, together with higher mean values ($\bar{x}$). Similarly, in two research series in the organic horizons of the spruce stand (F 1), equivalent relative standard deviations have been found.

Table 2. *Mean values ($\bar{x}$) and standard deviations ($s$) of the field Nmin content (mgNmin/100 g dried soil) of the mineral soil of the beech stand and the spruce stand. Sampling: beech = 4. 10. 1968, spruce = 18. 9. 1968*

| depth (cm) | spruce stand | | beech stand | | |
|---|---|---|---|---|---|
| | 6—12 | 14—20 | 0—6 | 6—12 | 14—20 |
| $n$ | 20 | 20 | 10 | 10 | 10 |
| $\bar{x}$ | 0.21 | 0.19 | 0.43 | 0.50 | 0.49 |
| $s$ | 0.05 | 0.03 | 0.11 | 0.07 | 0.10 |
| $s\%$ | 25 | 15 | 27 | 14 | 21 |

In the mineral soil the Nmin contents on the dry-weight basis are substantially lower than in the organic layers, but the degree of variation is less. Table 2 gives examples of Nmin content of mineral soils at 2 depths under the spruce and 3 under the beech stand. Samples were taken out of ten pits from one (B 1) or from two sides (F 1) of every pit. Each sample was analysed separately. As would be expected, the variation falls with increasing size of sample. For the periodic sampling the material was taken not only from one side of a pit, but from 4 sides of several pits, and mixed before analysis. Table 3 shows the results of analyses, when material from 4 sides of a pit or from 2 or 3 pits has been mixed. If we consider the material from one side of the pit a single sample, then the Nmin content of a mixed sample from 4 sides

13*

represents the mean value of 4 single values. Accordingly, a mixed sample from 2 pits would give a mean result of 8 single values, and a mixed sample from 3 pits a mean result of 12 single values. Consequently, the standard deviation of the Nmin contents of the mixed samples can be calculated from the known standard deviation of the single samples as "error of the mean value". Table 3 demonstrates that the standard

Table 3. *Mean values ($\bar{x}$) and standard deviations (s) of the field Nmin content (mg Nmin/100 g dried soil) of the mineral soil, using mixed samples from 1,2 or 3 pits (see text). Sampling: 18.9.1968. (anal. = analysed; calc. = calculated)*

|       | 1 pit anal. | calc.  | 2 pits anal. | calc.  | 3 pits anal. | calc.  |
|-------|-------------|--------|--------------|--------|--------------|--------|
| $n$   | 10          |        | 10           |        | 10           |        |
| $\bar{x}$ | 0.207   | 0.214* | 0.203        | 0.214* | 0.208        | 0.214* |
| $s$   | 0.027       | 0.027  | 0.012        | 0.019  | 0.016        | 0.016  |
| $s\%$ | 13.0        | 12.7   | 5.9          | 9.0    | 7.7          | 7.5    |

* Mean value from 20 single samples.

deviations, determined by analysis of mixed samples or by calculation from the standard deviation of single samples, are in good agreement. In one case the determined value is lower than the calculated one, probably because of the small number of samples.

### b) Vertical Distribution

It is evident from Table 1 that the various horizons of the organic layer differ considerably in their respective Nmin contents. As a rule the field Nmin content, based on dry weight, decreases in the following order: $F_1 > F_2 > L > H$ horizon. This succession is shown clearly by the mean values of the field Nmin content from all 33 periodic investigations (Fig. 1a).

Fig. 1a shows the mean values of the field Nmin content at the different depths of the mineral soil, too. As expected, the Nmin content decreases with increasing depth, but the differences are small.

However, these data, based on dry weight, do not give a true picture of the distribution of the mean Nmin reserve in soils. On converting the values to an area basis, i.e. allowing for the higher weight per volume and the greater depth of the mineral soil, it becomes evident that the Nmin reserve of the mineral soil is considerably greater than that of the organic layer. Fig. 1b shows the vertical distribution of the Nmin reserve.

In winter and spring the values are generally slightly higher, decreasing distinctly in late summer or autumn*. Table 4 shows, together with the mean values, the highest and lowest values in 1968, the distribution in the various horizons and the proportions found in the organic layer and the mineral soil of the beech stand.

---

* The alteration of the field Nmin contents in different horizons, and the alteration of net mineralization over the years will be described in a forthcoming publication.

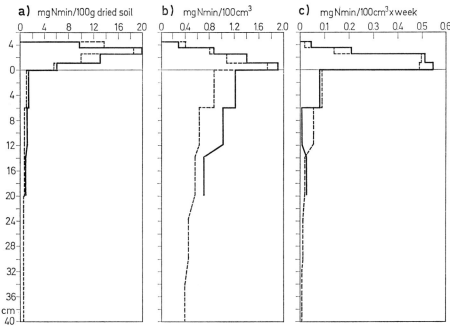

Fig. 1. Vertical distribution of field Nmin content and of net mineralization in the soil of the beech stand B 1. a) Yearly average of the field Nmin content, based on dry weight. b) Yearly average of the field Nmin content, based on soil volume. c) Yearly average of the net mineralization, based on soil volume. ——— = 1967; −−−−− = 1968. Re. b) and c): The length of the horizontal columns shows the field Nmin content or the mineralization in the different horizons under an area of 100 cm² for a depth of 1 cm. Multiplying by the depth of the respective horizon gives the content or the production of Nmin for the whole horizon

Table 4. *Field Nmin content (kg Nmin/ha) of the different horizons of the soil of the beech stand*

| Horizon | average 1967 | average 1968 | 23. 4. 1968 | 26. 8. 1968 |
|---|---|---|---|---|
| L | 0.3 | 0.4 | 0.4 | 0.1 |
| $F_1$ | 0.9 | 0.8 | 0.6 | 0.5 |
| $F_2$ | 2.1 | 1.6 | 2.2 | 0.6 |
| H | 1.9 | 1.7 | 1.9 | 2.9 |
| organic layer | 5.2 (= 21%) | 4.5 = 17% | 5.1 = 13% | 4.1 = 40% |
| $A_e$ 0— 6 cm | 7.3 | 5.2 | 6.4 | 1.3 |
| B 6—12 cm | 6.1 | 3.8 | 4.6 | 1.4 |
| B 12—14 cm[a] | 1.7 | 1.2 | 1.8 | 0.4 |
| B 14—20 cm | 4.2 | 2.8 | 6.0 | 0.8 |
| total (20 cm depth) | 24.5 | 17.5 | 23.9 | 8.0 |
| B 20—24 cm* | | 2.3 | 3.4 | 0.5 |
| B 24—30 cm | | 2.6 | 4.3 | 0.7 |
| B 30—34 cm* | | 1.6 | 2.8 | 0.4 |
| B 34—40 cm | | 2.2 | 4.1 | 0.7 |
| mineral soil | 19.3 (= 79%) | 21.7 = 83% | 33.4 = 87% | 6.2 = 60% |
| total (40 cm depth) | (24.5) | 26.2 | 38.5 | 10.3 |

[a] Mean value, calculated from the depths above and below.

## 2. Net Mineralization

### a) Significance of Differences between the Nmin Contents before and after 6 Weeks of Storage

The degree of exactness with which the field Nmin content of different horizons can be determined is restricted by their horizontal variability, whereas the exactness of the differences determined between the Nmin contents of mixed samples before and after 6 weeks of storage is mainly a question of method: the analytical error and the homogeneity of the mixing determine the variability of the results.

Mixing was done by hand, before storing parts of the mixed sample at the site. Better homogeneity could be achieved by sieving the material at this stage and not only before analysis, but then the material from the organic layer would be destroyed. This leads, as Zöttl (1966) has shown, to a significant alteration of the net mineralization. In mineral soils sieving can influence the net mineralization, too (Runge, 1965).

Organic material, mixed by hand, retains as certain heterogeneity, as is evident when the variability of analyses of different materials is compared. For instance, the series of analyses from one filtrate of material containing more than 10 mg $NH_4$-N or $NO_3$-N/100 g dry soil shows a relative standard deviation of 1.0—1.5%. When parts of a hand-mixed sample are taken, left at the site for 6 weeks and then analysed, the relative standard deviation is about 5—8%, provided the mean Nmin content is about 10 or more mg Nmin/100 g dry soil. The standard deviation decreases with decreasing Nmin content, but to a relatively smaller extent; consequently the relative standard deviation increases. However, except in winter, most differences were big enough to be significant.

Because of its better miscibility, the standard deviations in the mineral soil at comparable Nmin contents are generally smaller. But on the other hand the differences are smaller, too, especially in deeper layers. For that reason, differences often are not significant here.

### b) Vertical Distribution

Like the field Nmin content, the net mineralization shows a vertical distribution in the soil profile. Nmin production is most intense in the organic layer, fairly intense in a thin layer of mineral soil (0—6 cm), decreasing greatly below that (Fig. 1 c). Within the organic layer, the $F_2$ and H horizons have the highest Nmin production.

In the soil of the spruce stand the proportions are the same. Here, too, the Nmin production is concentrated mainly at the surface of the soil.

## IV. Discussion

### 1. Methodological Consequences

The results described have several methodological consequences for all investigations of net N mineralization:

a) The marked vertical distribution of field Nmin content and net mineralization points to the necessity for separate treatment of horizons or layers in every investigation of this kind. This becomes even more evident when not only mean yearly values (as in this publication) but the yearly progress of net mineralization are considered.

b) Because of the great variability of the Nmin content of single samples, it is only practicable to use mixed samples for periodically repeated investigations. Otherwise the number of single samples which would have to be analysed in order to provide satisfactory mean values would be excessive, particularly since the analyses have to be done both before and after storage.

The use of mixed samples has further consequences:

c) During periodic investigations, the variability of the field Nmin content of single samples (or of a certain number of mixed samples) should be examined on particular dates to test the validity of the mean values determined for the mixed samples.

d) In most cases the destruction of the natural soil structure by sampling and the following mixing will probably influence the net mineralization. Before the results are interpreted quantitatively, the magnitude of the "mixing effect" should be studied.

The results so far obtained on the mixing effect are not definite, but it can be said already that the disturbance of the natural soil structure, resulting from the sampling, has sometimes a stimulating effect on net mineralization.

Both examinations require large numbers of analyses to be carried out in a relatively short time, to avoid changes of the Nmin content in the filtrates. In this respect, the colorimetric methods were unsatisfactory. Once the steam-distillation method was proved suitable and much quicker, these examinations can be carried out now in more detail than was possible before.

The horizontal variability of the field Nmin content in the different horizons will be investigated with more samples at different times of the year and under different climatic conditions. Probably the relative standard deviation will sometimes be still higher than in the examples given.

## 2. Field Nmin Content

Up to now only a few special investigations have been made of the field Nmin content of soils, probably because it does not provide direct conclusions on the N nutrition of the vegetation. It is generally assumed that the reserve of exchangeable Nmin in the soils of natural or nearly natural sites is very low. In the natural N cycle there should be equilibrium between Nmin production and Nmin assimilation with only short-term fluctuations. Moreover, in many cases the N supply to the vegetation is deficient.

But the unexpectedly high Nmin content of the soils investigated in the Solling does not seems to be exceptional. STÖCKER (1968) has published similar or even higher results of analyses in the organic layer of mountain spruce forests.

Why is it that soils can permanently have such relatively high Nmin contents? Nmin production and Nmin loss from the soil (= Nmin uptake by the vegetation + Nmin losses by leaching + gaseous N loss) must be in equilibrium in these cases, too, as otherwise a permanent rise or fall would result. Consequently the high Nmin content can only be the result of a retardation in the Nmin loss, mainly in the Nmin uptake.

The most probable cause for this retardation is the limited number of absorbing roots in the soil. Nmin production takes place everywhere in the upper layers of the soil, Nmin uptake, on the contrary, only in a certain zone adjacent to the root tips. Accordingly, the Nmin content in the soil is determined on the one hand by the intensity of Nmin production, and on the other hand by the number of absorbing roots, by the intensity of root growth, and possibly by the transportation of Nmin to the roots. Varying proportions between these processes could cause the differences between different soils, the vertical distribution and the horizontal and seasonal variability.

Other explanations are possible, proceeding from the fact that high Nmin reserves have so far been observed mainly as $NH_4^+$ in soils with low pH values. This combination can be unfavourable for many species (Evers, 1964). But as these explanations seem to be less important, they will not be dealt with in detail.

### 3. Net Mineralization

The vertical distribution of net mineralization, described above, is characteristic of the humus form moder (Runge, 1965). The concentration of the net mineralization at the surface of the soil results from the general mineralization dynamics of moder soils. The organic material is very little intermixed with the mineral soil because an adequate soil fauna is lacking (see Babel, chapter V a). In the L horizon the C/N ratio is so great that at the beginning of decomposition nearly all the nitrogen present is used by the microorganisms. In the $F_1$ horizon the C/N ratio is narrowed to the extent that gross mineralization considerably exceeds microbial consumption. In the $F_2$ horizon the net mineralization, based on dry soil, is lower than in the $F_1$ horizon. Consequently the organic substances must be more stable here than in the $F_1$ horizon. This applies even more to the H horizon.

The Nmin production of the whole $F_2$ horizon and, at least at the beech site, of the H horizon too, is considerably greater than that of the $F_1$ horizon, due to the higher volume-weight and greater depth of the horizons. At the beech site the $F_2$ horizon supplies about 30% of the whole Nmin production of the soil. 50% is produced in the $F_2$ and H horizons, whose combined volume is only 7% of the whole space investigated. It is interesting in this connection that a marked concentration of roots is found in these two horizons.

The decrease in net mineralization within the mineral soil obviously depends on the decreasing humus content. In the deeper layers a seasonal change was observed between microbial Nmin assimilation (i.e. a decrease in the Nmin content of the samples during the storage) and net mineralization. Averaged over the year, net mineralization prevailed, but at 34—40 cm depth the surplus was only about 1% of the whole Nmin production of the site.

## V. Review of Further Results
### 1. Meadow Soils

Similar investigations to those reported above have been made in a meadow soil (W 2). Three plots have been compared: unfertilized, PK-fertilized and NPK-fertilized. PK fertilizing neither resulted in higher field Nmin contents nor did it distinctly change net mineralization, compared with the unfertilized soil. NPK-fertilizing led to a short-term increase in the field Nmin content, falling within a few days to the level of the other plots when the soil moisture was favourable. The net mineralization was raised significantly during the year, but this increase was restricted nearly exclusively to the upper 6 cm of the profile.

Interestingly, the progress of net mineralization over the year was basically similar in all plots in both years of the investigation, and was independent of the water content of the soils. After a pronounced maximum in spring, the net mineralization declined very much towards summer, when microbial Nmin assimilation temporarily exceeded gross mineralization. During summer and autumn the net mineralization was relatively low, with only small alterations. This corresponds to the productivity of the vegetation (see SPEIDEL and WEISS, chapter I) and to the growth behaviour of perennial grasses in general (RAPPE, 1966). The maximum in spring is a frequent phenomenon, probably due to an enrichment of easily mineralizable N compounds during the winter. The minimum in early summer is probably connected with the annual rhythm of root growth and root decay of the grasses.

## 2. Forest Soils

In the forest soils the annual course of field Nmin content and of net mineralization differs from that of meadow soil, but is similar in principle for both years of investigation. From the comparison between the changes in field Nmin content and net mineralization, a yearly course can be plotted for Nmin loss within single horizons or in the whole soil profile. This Nmin loss is largely determined by the Nmin uptake of the vegetation. At the beech site in both years the Nmin loss attained its maximum in late summer or autumn. This is in agreement with the results of older investigations by RAMANN and BAUER (1911) and KÜBLER (cit. after BÜSGEN, 1927): the beech shows its maximum N uptake in autumn.

The forest soils in the Solling have very low pH values, especially in the H horizon and at the surface of the mineral soil. In late summer values of about pH = 3.0 were measured in material from this horizon in water suspension. Nevertheless, a considerable part of the Nmin is found as $NO_3^-$; in summer, temporarily up to one third of the whole Nmin production. The $F_2$ and H horizons show the most intense nitrification. More $NO_3^-$ than $NH_4^+$ is produced during the vegetation period in the H horizon (the horizon with the lowest pH). These results indicate that even at very low pH, departing from the general rule, nitrification can be possible. This is true especially of organic layers, as was seen in earlier investigations (RUNGE, 1964). These conditions demand further investigation, as the $NH_4/NO_3$ ratio of the N nutrition is of importance for many plant species (EVERS, 1964; BOGNER, 1968; SCHLENKER, 1968). The question, which organisms perform the nitrification at these low pH values is being investigated by microbiologists within the Solling Project.

From the progress of net mineralization over the year, it is possible to calculate the annual Nmin production of the various horizons or of the soil as a whole. Quantitative data can be given when the investigations on the mixing effect are completed. These data will then have to be compared with the results of other projects that deal with the nitrogen turnover in ecosystems.

# References

Bartholomew, W. V., Clark, F. E.: Nitrogen transformation in soil in relation to rhizosphere microflora. 4. Int. Congr. Soil Sci. Amsterdam 1950.

Bogner, W.: Experimentelle Prüfung von Waldbodenpflanzen auf ihre Ansprüche an die Form der Stickstoff-Ernährung. Mitt. Ver. Forstl. Standortskunde u. Forstpflanzenzüchtung 18, 3 (1968).

Bremner, J. M., Edwards, A. P.: Determination and isotope-ratio analysis of different forms of nitrogen in soils: 1. Apparatus and procedure for distillation and determination of ammonium. Soil Sci. Amer. Proc. 29, 504 (1965).

— Keeney, D. R.: Determination and isotope-ratio analysis of different forms of nitrogen in soils: 3. Exchangeable ammonium, nitrate and nitrite by extraction-distillation methods. Soil Sci. Soc. Amer. Proc. 30, 577 (1966).

Büsgen, M.: Waldbäume, 3. Aufl. (ed. by E. Münch). Jena: G. Fischer 1927.

Conway, E. J.: Microdiffusion analysis and volumetric error. 5th ed. London: Crosby-Lockwood 1962.

Ehrhardt, F.: Untersuchungen über den Einfluß des Klimas auf die Stickstoffnachlieferung von Waldhumus in verschiedenen Höhenlagen der Tiroler Alpen. Diss. Staatswirtsch. Fak. Univ. München 1959.

Ellenberg, H.: Stickstoff als Standortsfaktor. Ber. dtsch. Botan. Ges. 77, 82 (1964).

Eno, F.: Nitrate production in the field by incubating the soil in polyethylene bags. Soil Sci. Soc. Amer. Proc. 24, 277 (1960).

Evers, F. H.: Die Bedeutung der Stickstoff-Form für Wachstum und Ernährung der Pflanzen, insbesondere der Waldbäume. Mitt. Ver. Forstl. Standortskunde u. Forstpflanzenzüchtung 14, 19 (1964).

Fawcett, J. K., Scott, E.: A rapid and precise method for the determination of urea. J. clin. Path. 13, 156 (1966).

Hesselman, H.: Studium über die Nitratbildung in natürlichen Böden und ihre Bedeutung in pflanzenökologischer Hinsicht. Medd. Stat. Skogsförs. Anst. 13/14, 297 (1917).

Jansson, S. L.: Tracer studies on nitrogen transformations in soil with special attention to mineralisation/immobilisation relationships. Kungl. Lantbr. Ann. 24, 101 (1958).

Ramann, E., Bauer, H.: Trockensubstanz, Stickstoff und Mineralstoffe von Baumarten während einer Vegetationsperiode. Jb. Wiss. Botan. 50, 67 (1911).

Rappe, G.: A year rhythm in production capacity of gramineous plants. B: III. Soil culture in photothermostats. Tests and determinations on soils and plants harvested. Oikos, Suppl. 7, (1966).

Runge, M.: Untersuchungen über die Mineralstickstoff-Nachlieferung an nordwestdeutschen Waldstandorten. Diss. Math.-Nat. Fak. Univ. Hamburg 1964; dito Flora 155, 353 (1965).

— Untersuchungen zur Mineralstickstoff-Nachlieferung am Standort. Flora 159, 233 (1970).

Schlenker, G.: Kulturversuche mit Waldbodenpflanzen bei abgestufter Azidität und variierter Stickstoff-Form. Oecol. Plant. 3, 7 (1968).

Stöcker, G.: Konzentration löslichen $NH_4$-N in organischen Horizonten naturnaher Berg-Fichtenwälder. Flora B 158, 41 (1968).

Yerly, M.: Ecologie comparée des prairies marécageuses dans les Préalpes de la Suisse occidentale. Veröff. Geobot. Inst. ETH, Stiftg. Rübel, Zürich 44 (1970).

Zöttl, H.: Bestimmung der Stickstoffmineralisation im Waldhumus durch den Brutversuch. Z. Pflanzenernähr., Düngung, Bodenkunde 81 (126), 35 (1958).

— Methodische Untersuchungen zu Bestimmung der Mineralstickstoffnachlieferung des Waldbodens. Forstw. Cbl. 79, 72 (1960a).

— Die Mineralstickstoffanlieferung in Fichten -und Kiefernbeständen Bayerns. Forstw. Cbl. 79, 221 (1960b).

— Einfluß der mechanischen Zerkleinerung von Streulagenproben auf Nmin-Anhäufung und $CO_2$-Produktion im Brutversuch. Plant and Soil 29, 2, 336 (1966).

# Y. Phenological Comparisons of the Forest Research Area in the Solling with Similar Forests in Other Mountain Ranges

F. K. HARTMANN

## I. Introduction

Phenological investigations examine the phases of plant growth throughout the year. At various sites the observed commencement dates of different growth phases in certain plants are studied to find their annual rhythm. They are then compared to provide indications of local differences in the climate.

By comparing the phenological data of growth phases in different years, information is also acquired on weather conditions in general and how these influence the development of plants.

Forest ecology in particular requires observation of the progress of flowering (generative phase) as well as of the development of the leaves of forest trees, shrubs and ground vegetation and finally, in late summer, the beginning of the decay (vegetative phase) and the ripening of fruit (again generative phase).

This can be done either by continuous observation of the commencement and duration of certain development phases of selected trees, shrubs or herbaceous plants, or by simultaneous statements on certain obervation days („Stichtage"). This method requires observers to be trained together in making the appropriate assessments, in order to produce comparable values in different observation areas.

Such a method has been in use by the Institut für Waldbaugrundlagen (Forest Ecology), University of Göttingen, since about 1955, and has also been applied to compare the Solling area with the other Weser hills, the Harz, Rhön and other mountain areas (see HARTMANN, SCHNELLE et al., 1970).

For the assessment of microrelief, the phenological behaviour of herbaceous plants on the forest floor has been worked out in some detail (see HARTMANN, VAN EIMERN, JAHN, 1959; HARTMANN, 1967; HARTMANN, SCHNELLE et al., 1970). Using the definitions "before spring", "early spring", "full spring", "early summer" and "summer" of SCHULZ one could add also other test plants to those named by him.

In acid soils there are relatively few plant species available as phenological indicators. Of these we chose *Oxalis*, *Vaccinium myrtillus* and, for "late spring" and "early summer", *Luzula albida*, because they are spread over many montains with acid soils. The number of test plants available on limestone or basalt is considerably greater than on acid soils.

# II. Methods

## 1. Interpretation of the Vegetative Phases in Spring and Early Summer

### a) General

On appointed days, the average state of development of a number of trees, shrubs and herbs growing at the observation site, is assessed and recorded on a form.

The development phases of trees are estimated separately on the upper canopy and in the middle and lower canopy. A definite phase of leaf development is assumed when at least half of the individuals have reached it (any variations or peculiarities should be mentioned).

### b) The Development Phases of Deciduous Trees

0     All buds still closed.
0.1   Buds expanded and opening, small green tips to be seen.
0.2   Leaves beginning to push out, the first small green leaf folds are visible.
0.3   Leaves have pushed out further, but not yet completely unfolded (e.g. beech), or in process of unfolding (e.g. oak, lime, maple, hornbeam, ash tree).
0.4   The part of the leaves which has unfolded amounts to 40% of the fully formed leaf surface.
0.5   ditto about 50%.
0.6   ditto about 60%.
0.7—1.0   ditto, until the full normal leaf surface is attained.

Individual variations (late developers, early developers) are to be noted. When upper and lower canopy have been estimated separately, the values are written above and below a horizontal line, e.g.

Bu $\frac{0,3}{0,5}$  in the upper canopy the leaves of the beech are pushing out vigorously, but have not yet unfolded; in the lower part of the crown and in the shorter trees, half of the normal leaf surface has already developed.

### c) Conifers

For conifers, the length of the so-called May shoot is given in cm (if possible also in tenths of the fully developed year shoot). Subsidiary branches are tested in comparison to the length of earlier shoots and to the length attained at the end of the vegetation period.

### d) Herbs and Grasses

0     leaves not yet unfolded or developed.
0.1—1.0   like deciduous trees.

For grasses, the height of the leaves is given in cm, and it is noted whether the stem has already extended, i.e. whether the shooting has begun ($= S$).

For ferns, it is estimated in tenths how far the frond is already "uncurled".

For seedlings (e.g. *Impatiens noli-tangere*) it is stated, as appropriate, how many pairs of leaves, apart from the cotyledons, have already developed.

## 2. Interpretation of the Generative Phases for All Categories of Plants

1     Flower buds not yet to be seen (phase BO of the German Meteorological Service).
2     Flower buds already visible, but still small and closed (onset of the blossom).

3a   Flower buds shortly about to open.
3b   Beginning of flowering, isolated blossoms at different places open (phase b of the Meteor. Serv.).
4    Full blossom, at least half of the buds opened (phase ab of the Meteor. Serv.).
5    Blossom falling, already partly with onset of the fruit.
6    Blossom completely over, with onset of the fruit.
7    Bearing fruit.
8    Fruits fully ripened.

### 3. Observation of the Autumn Aspect

The assessment of the various phases of the autumn aspect is based on the fruit ripeness, the discolouration and the falling-off of the leaves. The fruiting phases of trees, shrubs, herbs and grasses are determined according to the degree of maturity and colour variations of the fruits.

For the assessment of the duration of the vegetation period for wood growth, the onset of the first sign of discolouration and its speed of progress in deciduous trees is of great importance.* Particular attention should therefore be devoted to this in the different deciduous trees.

The discolouration of the leaves in trees can be followed in individual trees or in whole stands if only one species of tree is involved; otherwise the grades of discolouring and shades of colour of the individual trees are also noted.

Just as in spring the leaf development is estimated in tenths of the whole mass of leaves, so the autumn aspect should be assessed by the leaves still on the tree within the period of defoliation in tenths or percentages of the entire leaf mass. When doing this, the quantity of leaves already lying on the ground can be of considerable help.

## III. Phenological Comparison of Solling with Other Regions

Since 1955 such phenological surveys have been carried out by the author and his experienced collaborators in the Harz, the Rhön, the Pfälzer Wald and the Black-Forest (Schwarzwald), in connection with climate investigations. Some results of these surveys are given in HARTMANN, SCHNELLE et al. (1970). Numerous other observations have also been carried out in parts of the Upper Weser and Werra hills.

In the following we compare the areas of the IBP in the Solling mountains with other areas at similar or identical altitudes in the Harz, the hills of the Weser and Werra and the Rhön, arranged according to their phenological behaviour.

Fig. 1 shows the leaf development of the beech in phenologically different years. Fig. 2 shows the progress of flowering of ground plants *(Oxalis acetosella, Vaccinium myrtillus, Luzula albida)*.

In the Solling, 1967 was a year with average leaf development and 1968 a year with somewhat earlier development of beech (B 1). The flowering phase of ground vegetation was 1967 fairly early and 1968 less early.

The observed areas of the Harz cited for comparison in 1967 and 1968, include the Staufenberg near Zorge/South Harz, where we carried out exact climatic measurments in September 1953 and June 1954 (HARTMANN, v. EIMERN and JAHN, 1959),

---

* Compare the contributions of SCHOBER and SEIBT (chapter C) as well as that of LANGE and SCHULZE (chapter A) in this publication.

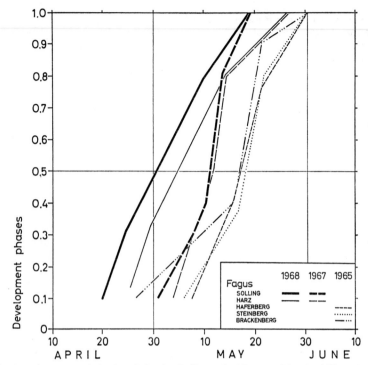

Fig. 1. Development of the beech in the Solling, the Harz and in the hill region of Upper Weser and Werra in different years

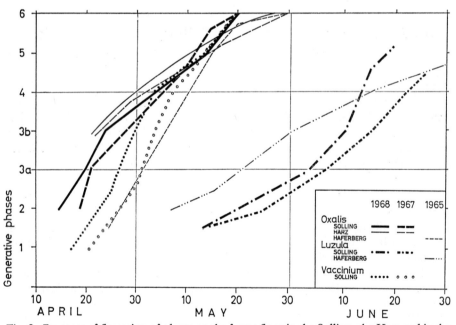

Fig. 2. Progress of flowering of plants on the forest floor in the Solling, the Harz and in the hill region of the Upper Weser and Werra in different years

at locations above 460 m (south slope); further the slopes of the large Wolfsbach Valley above Zorge at corresponding altitudes and the Nature Reserve Area "Dicke Tannen" near Hohegeiss. In addition, phenological observations on the north side of the Harz were made in the upper Ecker Valley and near Molkenhaus as well as at equivalent altitudes near the Rabenklippen and on the Woldsberg.

On the hills of the Upper Weser and Werra areas, 1965 was a year with late leaf development. Observation sites for these years were: Brackenberg, 450 m NN, basalt hill near Meensen (Kreis Hann. Münden); Steinberg, 540 m NN, basalt hill, montane/submontane; Haferberg 540 m NN red sandstone (Buntsandstein), montane/submontane, both in the northern Kaufunger Forest (Kreis Hann. Münden).

In comparable altitudes of the Rhön mountains, the phenological development in 1967 and 1968 was quite similar to that of the Solling research area.

## IV. Comparison of Phenological Development with Photosynthetic Activity

In regard to the causative relationships between changing atmospheric conditions and phenological development, it should be remembered that individual weather elements are able to strengthen or weaken one another. Yet it may be considered an established fact that the times when the leaves and blossoms appear are substantially influenced by temperature.

Studies by SCHULZE (1970) on tree 68 in the older stand of beech (B 1) in 1968 yield valuable comparative phenological data.

SCHULZE found the following phases to be quite differently developed, depending on whether sun leaves or shade leaves were being considered:
a phase of increasing photosynthesis
> in sun leaves until June 20, 1968;
> in shade leaves until August 28, 1968;
maximal assimilation
> by the sun leaves: mid-June until the end of July;
> by the shade leaves in September.

The autumn phase also occurred at different times in sun and shade leaves. The sun leaves in the fully illuminated crowns of the tree observed by SCHULZE began to change slightly in colour as early as the 6th—9th of August; namely in a hot period with hourly temperature means of up to 23.5° C. $CO_2$ assimilation by the shade-leaves, however, began to decrease as late as the 17th—20th of October, 1968, but soon ceased completely. The yellowing and failing of the shade leaves likewise began very abruptly.

SCHULZE observed heavy blossoming of his test tree on April 29, 1968. According to our observations, in the area as a whole, blossoming began somewhat earlier.

## References

HARTMANN, F. K.: Zur Beurteilung wachstumsentscheidender Standortfaktoren nach soziologischen-ökologischen Artengruppen — Möglichkeiten, Grenzen. Vortrag auf dem XIV. IUFRO-Kongreß, Sektion 21, München 1967.

HARTMANN, F. K., EIMERN, J. VAN, JAHN, G.: Untersuchungen reliefbedingter kleinklimatischer Fragen in Geländequerschnitten der hochmontanen und montanen Stufe des Mittel- und Südharzes. Ber. dtsch. Wetterdienst, (50) (1959).

— SCHNELLE, F. et al.: Klimagrundlagen natürlicher Höhenstufen und ihrer Waldgesellschaften in deutschen Mittelgebirgen. Reihe: Ökologie der Wälder und Landschaften, ed. by F. K. HARTMANN, Bd. 4. Stuttgart: G. Fischer-Verlag 1970.

— JAHN, G., AMELUNG, E., BOSSE, J., MÜLLER, U., STREITZ, H.: Phänologische Beobachtungen und Teilauswertungen derselben aus Harz, Rhön, Nordschwarzwald, Pfälzer Wald und Weserbergland; unpublished, 1958—1965.

SCHNELLE, F.: Über die Abhängigkeit der Laubverfärbung von der Temperatur. Ber. dtsch. Wetterdienst **38**, 227—228 (1952).

SCHULZ, L.: Wie der Frühling in den Harz einzieht. Z. Naturlehre u. Naturkunde. Stuttgart: G. Fischer-Verlag.

SCHULZE, E. D.: Der $CO_2$-Gaswechsel der Buche (*Fagus silvatica* L.) in Abhängigkeit von den Klimafaktoren im Freiland. Flora **159**, 177—232 (1970).

# Z. Results of a Grassland Mapping in the High Solling

B. SPEIDEL

## I. Grassland Communities Mapped

In order to be able to assess the range of validity of the studies on the grassland sample plots W 1 and 2, a vegetation mapping was carried out on about 400 ha of grassland around Silberborn and Neuhaus. The following communities were found to be present:

% of the mapped area:

1. *Lolio-Cynosuretum*
   1.1. sub-association of *Dactylis glomerata*
   1.2. typical sub-association                         } 2%
   1.3. sub-association of *Alopecurus geniculatus*

2. *Trisetetum flavescentis hercynicum*
   2.1. typical sub-association
      2.1.1. typical facies                  4
      2.1.2. *Festuca rubra* facies                     } 25%
          2.1.2.1. typical variant     20
          2.1.2.2. variant of *Nardus stricta*   1
   2.2. sub-association of *Polygonum bistorta*
      2.2.1. typical facies                  4
      2.2.2. *Festuca rubra* facies                     } 50%
          2.2.2.1. typical variant     45
          2.2.2.2. variant of *Nardus stricta*   1

3. *Nardetum strictae*                                   } 1%

4. *Polygonum bistorta* community
   4.1. typical facies
   4.2. *Festuca rubra* facies                     } 7%
      4.2.1. typical variant
      4.2.2. variant of *Nardus stricta*

5. *Juncus effusus* community
   5.1. typical facies                        } 7%
   5.2. *Carex fusca* facies

6. *Molinia coerulea* community

7. *Juncetum acutiflori*                              } 8%

8. *Carex rostrata* stands

9. *Phalaris arundinacea* stands

Of these communities the *Lolio-Cynosuretum*, the *Trisetetum* and the *Polygonum bistorta* community are regularly utilized. The *Nardetum* and the *Juncus effusus* community are for various reasons utilized only occasionally, while the *Molinia coerulea* community, the *Juncetum acutiflori*, the *Carex rostrata* stands and the *Phalaris arundinacea* stands are actually unutilized wet heathland.

## II. Applicability Range of the Production Studies

In the High Solling area the prevalent system of farming is that of scattered peasant smallholdings. The one exception is the 90 ha area which is occupied by a stud farm. This land is much more intensively utilized and also better fertilized than the normal farmland, and it contains only *Lolio-Cynosuretum*. It can be readily seen that this is not typical of the High Solling from the fact that this community occupies only 2% of the total area mapped except for the stud farm. If the validity of the results gained on the sample area is to be established, then this stud farm area must be omitted, and only the grassland farmed in the usual way may be taken into consideration.

The stand on the sample area is in accordance with the typical variant of the *Festuca rubra* facies within the typical sub-association of the *Trisetetum* (2.1.2.1.). Until the studies were begun the sample area was extensively utilized as grassland in the manner usual in this locality.

The *Trisetetum flavescentis* is characterized by the species *Alchemilla vulgaris*, *Phyteuma spicatum* and *Lathyrus montanus*, as well as by the characteristic species of the fertilized meadows belonging to the order *Arrhenatheretalia* (*Trisetum flavescens*, *Dactylis glomerata*, *Heracleum sphondylium*, *Pimpinella major* etc.). The typical sub-association (with no further differential species) indicates a normally drained soil, and the sub-association of *Polygonum bistorta* (with the differential species *Polygonum bistorta*, *Cirsium palustre*, *Deschampsia caespitosa* and other species belonging to the order *Molinietalia*) indicates a slightly periodically moist soil. In both sub-associations, *Festuca rubra* is dominant where the meadow is fertilized regularly.

For the assessment of the differences in the yield of the whole *Trisetetum*, community classification according to nutritional aspects is far more suitable than one according to soil moisture. Within the variability range of the whole association the typical facies of both sub-associations establishes itself on sites that are rich in nutrients and alkali, and therefore produces the highest yields. On rather less well-provided soils the typical variant of the *Festuca rubra* facies is found, with a correspondingly lower yield, while the *Nardus* variant is always situated on the poorest sites within the *Trisetetum*, so that compared with the typical variant its yield is even lower.

Depending upon the same management and fertilization intensity, the two *Festuca rubra* facies (2.1.2.1. and 2.2.2.1) are very similar and might even be regarded as one single grassland type. Together they occupy about 65% of the mapped area. The conclusion can thus be drawn that the sample plots W 1 and W 2 are representative of about two thirds of the grassland in the High Solling. The results obtained can therefore be regarded as characteristic of the area.

## References

Braun-Blanquet, J.: Pflanzensoziologie, 3rd ed. Wien-New York: Springer 1964.
Hundt, R.: Die Bergwiesen des Harzes, Thüringer Waldes und Erzgebirges. Jena: G. Fischer 1964.
Meisel, K.: Zur Gliederung und Ökologie der Wiesen im nordwestdeutschen Flachland. Schriftenreihe Vegetationskunde 4 (1969).
Oberdorfer, E.: Süddeutsche Pflanzengesellschaften. Jena: G. Fischer 1957.

# Subject Index